How to Lie with Numbers, Stats & Graphs

Copyright

How to Lie with Numbers, Stats & Graphs

By Lee Baker

Copyright 2020 Lee Baker

CSI Publishing

Part 1:
Truth, Lies & Statistics

or

How to Lie with Statistics

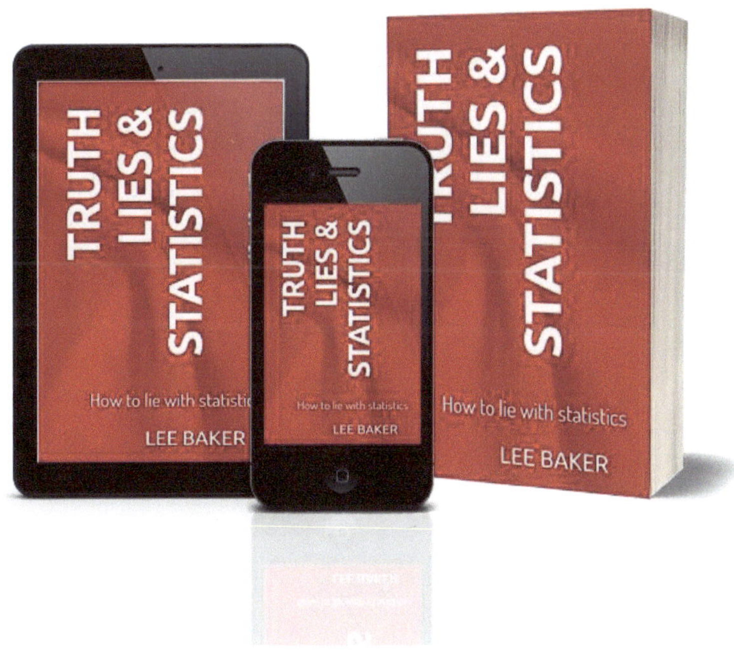

iii

Contents

The Uncomfortable Truth About The Truth

In the immediate aftermath of Donald Trump's inauguration as President of the United States, it was claimed angrily and repeatedly that the size of his inauguration crowd was the "largest audience to ever witness an inauguration – period…". White House staff backed this up by quoting the numbers of people that rode on the Washington DC Metro on that day and compared it to the figures for Barack Obama's inauguration.

What more proof do you need?

Well, for starters, we need these numbers to be quoted *correctly* (they weren't). The official DC Metro ridership figures show that Trump's inauguration crowd was actually the smallest of the last four inaugurations.

So, was it a lie? Well, it certainly wasn't the truth, the whole truth and nothing but the truth.

You see, it is easy to fall into the trap of thinking that truth and lies are binary. If it's not the truth, it must be a lie. If it's not a lie, it must be the truth.

Not so. There is a grey area in-between that we all fall into from time to time, whether by accident or otherwise.

I'm sure you've heard of such terms as half-truths, partial truths, preferred truths, uncomfortable truths and alternative facts. Similarly white lies, fabrication, exaggeration, bias and deception. These all fall into that grey area.

Statistics also falls into that grey area.

Before I get an avalanche of emails decrying me as the spawn of the devil, let me explain.

Data doesn't lie. People do. If your data is biased, it is because it has been sampled incorrectly or you asked the wrong question (whether deliberately or otherwise).

Statistics, on the other hand, does lie. When analysing data it is rarely the case that there is one correct approach and all other ways are wrong. There are often many ways of analysing the data, some of which are more appropriate than others, and these different approaches usually give different answers. Take as an example 'the average' – the central point of your data. There are over a dozen different ways of calculating it. Take a sample of data and work out all the different averages. Which one of them is correct? Most likely none of them. Some of them will be closer to the true central point of the data than others, but it is extremely rare that statistics gives you the truth and nothing but the truth.

By careful selection of a particular statistical method you can get a result that is close to the truth or very far from it, as is your wont. And this is what I mean by saying that statistics lies. It rarely tells the entire truth and you can make it as close to, or as far from, the truth as you like. If you've spent enough time around statistics, you'll know that you can make your data say pretty much whatever you want it to, if you're so inclined.

Of course, *good* scientists, researchers and statisticians wouldn't dream of doing such a thing, would they? Or would they...

Have you ever heard the phrase '9 out of 10 cats prefer...'? Of course you have. After all, they are a multi-billion dollar company who has commissioned a hugely expensive advertising campaign to persuade us to buy *their* cat food rather than that of their rivals. And yet seemingly they can't afford to have more than ten cats in their trial. There's definitely something fishy here, and it's not just the flavours of their cat food.

If the researcher works for a commercial organisation, the results of their analyses will always be subject to question because – even if their genuine, truthful and unbiased results are strong – artificially inflating the results can lead to a greater volume of sales. The company has something to gain.

Similarly if you work for a contract research outfit that is being paid by a company to do some research and analysis. On one hand if you produce a biased study you will leave the integrity of your company open to debate, but on the other, a future contract with the paying client might be at stake if the results don't sparkle quite as brightly as they might. "More than 80% of dentists recommend our toothpaste", claimed one advertising campaign. Does that mean that fewer than 20% of dentists recommend all the other toothpastes combined? Not necessarily, but that's the clear implication. The data collected may have been correct, but was the survey designed to mislead? Almost certainly!

You don't need to be creative with your choice of statistical analysis to deceive, though. Oh, no, we humans are hard-wired to look for patterns. Just show us a pretty little pattern in the data and we'll make up our own story to explain it. Like how global warming has increased as the number of pirates has decreased. The obvious inference is that the lack of pirates caused the planet to warm. It's true, honest, and there's a new religion that's sprung up around it. More on that story later…

Better still, we might even be able to use statistics to 'prove' that the pattern is a real one, like the correlation between autism and organic food sales. Does organic food cause autism, or is it the other way round? Similarly with the correlation between imported Mexican lemons and deaths on US highways, implying that Mexican lemons are killing Americans. There surely *must* be something in these because the correlations are highly significant.

I hope you can see that it is not possible to tell the whole, unvarnished truth all the time. You can't even tell it most of the time. In fact, you might not be able to tell it at any time, but if you are honest about all sources of potential bias and weaknesses in your study methods, you will come out of it with more credit. And that's never a bad thing.

On the other hand, if you're comfortable with half truths and alternative facts – and even prefer them as a vehicle for profit or career enhancement – then this book will help you find lots of ways to tell untruths with your data.

A word of caution though: this book is not written for you. It's written to help good, conscientious researchers spot when you're trying to pull a fast one. For every conman there has to be a good, honest man ready to point an accusing finger.

After all, as the saying goes – evil flourishes when good men do nothing...

Are You Biased?

The short answer to this question, as we have already established, is Yes.

The long answer is still Yes, but with the caveat that biases may be understood, controlled for and minimised – if indeed that is your intention – but not eliminated entirely.

Take the example of the airline industry. The United States airline company Boeing reported over 30,000 passenger fatalities in their 2016 annual Statistical Summary of Commercial Jet Airplane Accidents report. Wow – that is a huge number of deaths of American citizens in a single year.

Clearly it's not safe to fly.

Except that it's not true. What is true is that Boeing reported over 30,000 deaths in their 2016 annual report. By carefully choosing my words, I accidentally on purpose made it look like all these deaths occurred in 2016. They didn't. In fact Boeing reported that these were the cumulative deaths from 1959 to 2016, and not just in the US, but worldwide. The average figure (for what it's worth – it's ridiculously easy to lie with averages, as we shall see later) is just 526 worldwide airline passenger deaths per year, and the annual accident rate per million departures has declined annually from 50 in 1959 to around 1 in 2016.

Clearly it's safe to fly – and becoming increasingly so.

You see, I told the truth, but I did not tell the *whole* truth. I deliberately deceived by using a number of different biasing techniques. I was highly selective in reporting only the facts I wanted you to see and added in some other stuff to artificially inflate the figure in the reader's mind. And I did this without actually telling a lie.

Never mind that Boeing's numbers were recorded, tallied, analysed, calculated, graphed and nicely reported, if you have a mind to misrepresent the data it is not difficult to do so. Newspapers and politicians do it all the time and they are constantly being rewarded for it (more newspaper sales, more votes).

Bias is a type of error in which a measure or observation is *systematically* different from the 'truth' (whatever that is). It can affect any stage of the research process, from your literature review to measuring and recording your data, from analysing and interpreting your results to publishing them in a journal, thesis, report or newspaper article.

Imagine you sent out questionnaires to a thousand households, asking a single, simple question: Do you like to answer questionnaires?

I hope it's obvious to you that those people who do enjoy questionnaires will answer that they do, while those that do not will file the paperwork carefully in the nearest waste-paper bin. The former group of people will have **self-included** themselves in the study, while the latter will have **self-omitted**. Any analyses, results and interpretations based upon these flawed data will have almost zero validity. This may be a very extreme example, but **selection bias** like this is deliberately inserted into data every day by unscrupulous commercial organisations that are more interested in selling their wares to you than presenting the truth.

Like the dieting programme that rejects the data of all those that drop out of the trial — most of those that drop out likely do so because it wasn't working, whereas those that successfully lose weight are likely to stay in. Report your results, data be damned!

It is not just a simple matter of saying 'I'm not biased, so therefore I won't bias my study', though – bias may be **conscious** (deliberate) or **unconscious** (accidental, careless or negligent), and there are literally hundreds of ways in which you can bias your research.

And ignorance is no excuse.

If you're not experienced in data gathering, a more experienced practitioner will easily spot bias in your processes and tear your study to shreds, rendering your 3-year study worthless. Not a nice feeling!

And just because you've made calculations to sixteen decimal places, don't think for one second that your numbers are in any way valid. A calculation based on poor assumptions and flawed measurements, however accurate it may seem, is just as guilty of being biased as if you'd just pulled the number out of a hat.

Incidentally, a few years back I did some analysis for a colleague who wanted me to check some of his results. I shall call him Paco, for no other reason than that is not his name. Despite having precisely the same data and having used the same analysis methods, I was unable to verify some of Paco's results. When I asked why some were statistically significant in his analyses where they were not in mine, he replied 'well, if I exclude these 2 patients from the analysis the p-value becomes significant'. On enquiring why those patients should be excluded from the analysis, he said 'because when I do, the p-value becomes significant...'.

Oh dear – my good friend Paco was guilty of omission bias. He didn't make things any better for himself though; he was quite unrepentant, insisting that there was a very good scientific reason why he should exclude these patients from the study, but that he just hasn't found it yet!

Another way of adding bias into your study is by consistently understating or overstating a particular measure. If measurement errors are *systematically* biased you will be guilty of **measurement bias**, also known as **observational bias**.

You might spot this at unscrupulous weight loss clubs where they 'fix' the scales at first weighing to read a higher weight, then alter the scales to deliberately record a lower weight at subsequent weigh-ins. This ensures small but consistent weight loss is recorded, keeping the happy customer coming back for more.

A real-world example of measurement bias came in 1628, when the Swedish ship, Vasa, sank less than a mile into her maiden voyage, culminating in the deaths of 30 people. A recent investigation into the sinking of what was considered to be the most powerful warship in the world has discovered that the ship is thinner on the port side than the starboard side. Apparently, the workers on the starboard side used rulers that were calibrated in Swedish feet (12 inches), while workers on the port side used rulers calibrated in Amsterdam feet (11 inches).

Of course, in these days of great computing power and automated software programs these kinds of mistakes just don't happen. Or do they? A catastrophic and very expensive example of measurement bias came in 1999, as the Mars Orbiter was lost as it travelled too close to the planet's atmosphere. An investigation said the cause of the loss of the $125m probe was because the NASA team used metric units while a contractor used Imperial measurements. In other words, despite an abundance of space in that region of the solar system, the probe missed and crashed into Mars instead. Oops...

I particularly enjoy the story of the British rock band Black Sabbath, who, in 1983 had a replica of Stonehenge made for their stage show. Unfortunately, it was so big that it got in the way of the band and very few of the replica stones would fit on the stage. The legend goes that there was a mix-up between metres and feet, so the entire structure was three times bigger than it should have been. This was parodied the following year in the riotous This is Spinal Tap mockumentary, where the rock group ordered a model of Stonehenge for their stage show, but the note written on a napkin said 18'' (18 inches) rather than 18' (18 feet).

When it comes to reporting results, you can bias the outcome by highlighting favourable evidence that seems to confirm a particular position while suppressing evidence to the contrary. This is known as **cherry picking**, also known as **suppressing evidence**, and may be intentional or unintentional. In public debate, though, and particularly in politics, it is rarely unintentional.

After Donald Trump's inauguration as President of the United States, the White House Press Secretary, Sean Spicer, held a press briefing and angrily accused the media of deliberately underestimating the size of the inauguration crowd, stating that it was the "largest audience to ever witness an inauguration – period...". To back up his claims, Spicer (who was subsequently – and mercilessly – parodied by comedienne Melissa McCarthy on the US comedy show Saturday Night Live) claimed that 420,000 people rode the Washington DC Metro on inauguration day in 2017 compared to 317,000 in 2013. This statement led to a series of highly public (and, to the casual observer, highly amusing) arguments between the press and the White House staff about the meaning of the words 'truth' and 'facts'.

When Spicer quoted the number 317,000, he was attributing it to Barack Obama's 2013 inauguration, and it was a correct number, although it was **cherry picked**. The 317,000 referred to the number of people that travelled on the DC Metro in *the morning* of Obama's inauguration. The whole day figure was 782,000.

Conversely, no one knows where the figure of 420,000 for Trump's inauguration came from. The official figures for the ridership of the DC Metro were 193,000 in the morning and 570,000 for the whole day – the lowest figures of each of the previous four inaugurations!

In an interview the following day, Trump's campaign strategist Kellyanne Conway defended Spicer's statements, saying "Sean Spicer gave **alternative facts** to these claims".

Prior to Trump's inauguration, the term 'alternative facts' meant something different. It was generally understood to be 'OK, so you've shown me your facts, now let me show you mine', and the two sets of alternative facts (all of which were, actually, factual) were examined. The world now sees 'alternative facts' as a synonym for **deliberate falsehoods** or **intentional lies**.

For those interested in spotting bias in news reports and research studies, it is not enough simply to inspect the numbers being presented. It is useful to dig deeper and ask questions such as Who is the author? What do they have to gain? Who sponsored or commissioned the work? Has the author diligently highlighted potential sources of bias in their report? If not, why not? It is only by asking such questions that you can reveal the motives behind the author and come to your own conclusion about the validity of their arguments.

The Average Human Being Has Only One Testicle

A physicist, an engineer and a statistician on a hunting trip spot a deer in a clearing. The physicist shoulders his rifle, takes aim and misses the deer by 5 feet to the left. The engineer grabs the rifle and says, "You forgot to account for the wind". He takes aim and misses the deer by 5 feet to the right. The statistician suddenly jumps up into the air, shouting excitedly "We got it, we got it!".

When it comes to misdirection there is no better way than using averages. The term 'average' is used an awful lot in the popular press – and even appears far too often in respected peer-reviewed journals. You see, there isn't a single meaning when it comes to the word 'average', it can mean any of a dozen different things, all of which could be very different.

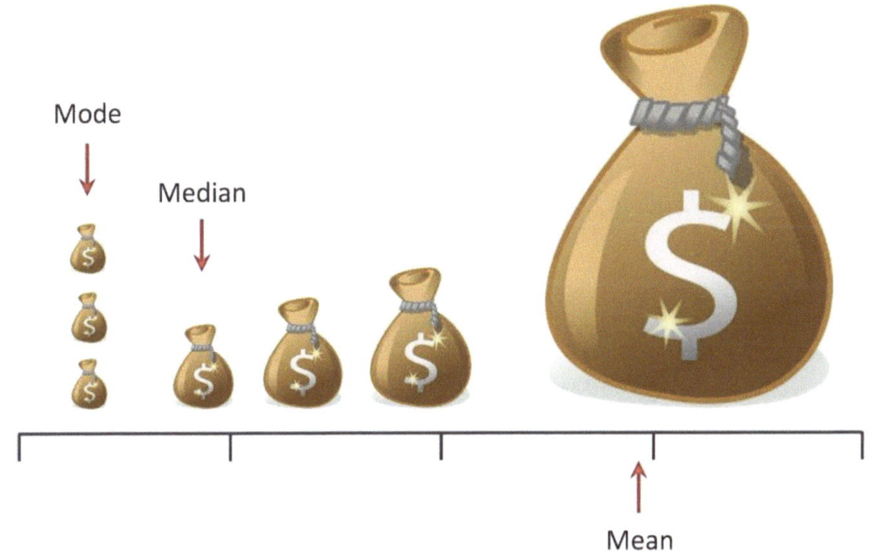

In general, when someone uses the word 'average' they mean the middle value. Let's look at an example to see how different measures of average (mean, median and mode) can give wildly varying answers.

Let's say that there are ten people that live on your street and that, on average, everyone is a millionaire. Happy days! Now what if I told you that three are out of work and are flat broke, six are employed on normal wages of about £20-30k, and the legendary rock musician Ozzie von Hendrix, who earned £10 million last year, lives in the mansion on the cliff at the end of the street. That would rather change the whole complexion of the neighbourhood, wouldn't it?

You see, it all depends which measure of average you choose to use. If you're a real estate agent and you're trying to sell a house in the street, it will do your cause the power of good to tell prospective buyers that this is a really good area with an average income of over a million quid (the mean). On the other hand, you could argue to the local taxes commission that the residents cannot afford any rate rises because the average resident for this street is unemployed and hasn't earned a bean in the past year (the mode in these data is £0).

And yet the numbers don't lie – you could argue that on average you are all millionaires (mean), earn a standard living of £25k (median) and at the same time you are all broke (mode). This is why you should get a very nervous tick whenever you hear the word 'average' without an explanation of which specific measure is used.

There is a specific case, though, when mean, median and mode are all the same, and that is when your data are normally distributed. It is useful to understand when data are not likely to fit a normal distribution when trying to figure out if someone is trying to pull a fast one. Heights of men are normal, as are blood pressure measurements or IQ scores. Beware, though, the data distributions that might sound like they would be normal, but in fact hide two or more distinct distributions, such as annual salaries.

It is not difficult to rationalise annual salaries in a company as being normally distributed – the janitors and cleaners are the lowliest paid, the managers are more highly paid and the bulk of the workforce have a wage somewhere in between. Sounds normal to me. Ah, yes, but what about the directors and executives? Typically, they have an entirely different pay scale. If you were to graph all the salaries with income on the x-axis and frequency on the y-axis (the number of workers in a particular pay band) there would be a large hump on the left representing the wages of the employees, a gap, and then a smaller hump on the right corresponding to the wages of the executives. If one were to include all wages in a single 'company average salary', the executive salaries would skew the results of the mean wages to the right, suggesting a higher 'average' salary. The median salary, though, would hardly be affected by the executive wages.

If the intention of the analyst was to try to downplay the average company wages, perhaps for some sort of tax audit, she might omit the executive wages (remember selection bias from above?) and then use the median. Alternatively, to exaggerate the average salary, perhaps for recruitment purposes, she would include executive salaries and use the mean. Better still, make sure that all executive bonuses, dividends and payments-in-kind are included too – that would make for an even higher average!

Whenever the word 'average' is used, we should be sceptical – it is very often used to mislead. Like the man who drowned in a pool of water, the average depth of which was one inch. Or that the average human being has only one testicle.

We should also spot when a report specifies which measure of average is being used for *some* figures but not all: 'Our survey said that median age was 27 years old and average income was £32k'. The attempt to hide a rogue 'average' income in there amongst a 'median' age should not fool the wary, and questions should be asked as to the validity and motives of the report.

Give Ps A Chance

A physicist, a chemist and a statistician spot a fire in the waste paper basket. The physicist decides that to put out the fire they must cool down the materials until the temperature is lower than the ignition temperature. The chemist disagrees, deciding that they should eliminate one of the reactants by cutting off the oxygen supply. With the debate raging, the physicist and chemist are alarmed to see the statistician running around starting fires in all the other waste paper baskets. "What are you doing?" they both scream, to which the statistician replies, "Trying to get an adequate sample size...".

What does it mean when a survey finds that "66.7% of people agree that..."? You can be pretty sure that there were only three people in the survey and one of them disagreed. What if one person in the study changed their mind? Well, then, either 100% or 33.3% agreed. These percentage swings are huge, and – I'm sure you will agree – the survey has almost no credibility.

The number of participants in a poll, study or survey, then, is critically important in determining its validity. We call this the **sample size**.

If you're reading this book then you will likely have at least a nodding acquaintance with statistics and will understand why we need a sample – it's because most of the time it's practically, ethically or economically impractical to poll the entire population. Instead, we take a small representative sample and then scale up to infer the results of the whole population. Ah, yes, but how small? There is a whole branch of statistics dedicated to calculating sample sizes in all sorts of situations, and there is just as much art in the calculations as science.

What can be said with certainty, though, is that the larger the sample size the higher your confidence in the results – though the results might not be what you wish to find. If you're a smart, conscientious medical researcher whose job it is to save lives and whatnot, then you will try to have as large a sample size as your ethics committee will allow. Conversely, if you're a researcher who works for a company whose job it is to sell as much product as possible, then your main role might be to show how your product is 'better' or 'preferred' over rival products, even if it isn't. Not to worry – you have statistics and small sample sizes on your side!

This reminds me of a piece of research that led to Dr Andrew Wakefield being struck off the UK medical register for his 1998 research paper (amongst other things) that claimed there was a link between the measles, mumps and rubella (MMR) vaccine and the appearance of autism in children, which has since been shown to be fraudulent. The paper was reported widely in the media and panic ensued as millions of parents across the US, UK and beyond refused to have their children vaccinated with the MMR vaccine. After vaccination rates dropped, incidences of two of the three diseases increased greatly. As a direct result, many children who were not immunised became very ill, some were permanently injured and there were even deaths reported across the globe.

During an investigation into the study, all sorts of unethical and unauthorised shenanigans were uncovered, but there were two things in particular that really stood out for me. The first was that the paper had a sample size of just 12 children. As we've already established, small sample sizes are the playground of the unscrupulous, and any medical study that has only 12 samples should be taken with a pinch – correction – a huge cartful of salt. The second was that the paper was published in the Lancet – one of the most prestigious peer-reviewed publications in the world. Really. They should have known better. To their credit, the Lancet eventually retracted the paper claiming that it was "utterly false" and that the journal had been "deceived".

One of the big problems with this research was that with only 12 patients it was very easy to 'turn a blind eye' to a piece of evidence or 'tweak' the odd bit of data here and there and – hey presto – a highly significant result. It would have been much harder to deceive the research community if the sample size had been much higher.

OK, take a deep breath, count to ten and we'll continue with a lighter subject…

The bottom-line about sample sizes is that the larger the sample the higher the confidence in the results. Sure, but how do we measure this confidence?

Many years ago, the British statistician Ronald Aylmer Fisher invented the p-value. Well, actually, that is disputed, but I'm not getting into it. Fisher, at the very least *popularised* the p-value as a measure of confidence about a particular hypothesis test. We all know how to do this; you make a hypothesis that something is *not* true (the null hypothesis) and then try to disprove it. The p-value, a number between 0 and 1 tells us the probability that the null hypothesis is correct. If the p-value is large (greater than 0.05), there is strong evidence that the null hypothesis is correct and our confidence of a significant result is low. On the other hand, if the p-value is small (smaller than 0.05), there is weak evidence to support the null hypothesis and we can be confident in *rejecting the null hypothesis* and concluding that there is evidence of a significant result.

But where did this arbitrary figure of 0.05 come from? This came directly from Fisher himself, who said that:

…it is convenient to draw the line at about the level at which we can say: "Either there is something in the treatment, or a coincidence has occurred such as does not occur more than once in twenty trials"…

And of course, 1 in 20 expressed as a decimal is 0.05.

In seven years as a statistical consultant in one of the largest teaching hospitals in Europe, not a day went by when I did not receive a dataset with the instructions 'I need a p-value for my study'.

Since Fisher introduced the p-value in 1925 it has become one of the cornerstones of statistical analysis, and any study that does not have a p-value attached to it is deemed worthless.

So why is it that seemingly important numbers as "our survey found a 57% increase…" is not accompanied by a p-value? Do they have something to hide? You can bet your sweet tush that they do!

Take the example of the cat food commercial, where they claim, "9 out of 10 cats prefer MiaoMix cat food". To what? Budgie seed? And how many times did they repeat the experiment of putting cats in front of two bowls of food and waiting until 9 out of 10 went to the bowl that they wanted them to go to? If they had reported that '90 out of 100 cats preferred…' it would be much more believable, but it is very much easier to get 9 out of 10 cats to go to the 'right' bowl than 90 out of a 100, especially when you spray the 'wrong' bowl with cat repellent. You see, sample size matters, and so does the p-value. Where is the p-value? Nowhere to be seen! How many repeats did they conduct on the experiment? No idea – they won't tell us. And just where did they file the *other* results from all these repeat experiments? To quote the late, lamented Douglas Adams: They were on public display in the display department beneath the cellar in the bottom of a locked filing cabinet stuck in a disused lavatory with a sign on the door saying 'Beware of the Leopard'…

So the next time you see a report, paper, thesis or whatever, take the time to check out the sample size and ask yourself if the sample is large enough to permit any reliable conclusion, then follow up with the question 'what is the probability that this result is true?'. Look for whether a p-value is reported. If the study is statistically significant, you can bet your mother's top lip that the authors will include a p-value and anything else that would add weight to their assertions. If they are not there, be very, very suspicious.

The Confidence Trickster

"Look at the difference in heights on this bar chart," says the researcher, "there *must* be a significant difference between them – this is twice the height of the control sample."

"Ah yes," says the statistician "but where is your confidence about the result?"

"Oh, I'm very confident about the result…" explains the researcher.

Statistician face palms, exclaims "d'oh!".

Measure the heights of 100 randomly chosen men, and calculate that the mean height is 175cm. If you repeat the experiment (with 100 *different* men, obviously), will you get precisely the same answer as you did first time? If you get a different answer, how do you know which one is correct? Are either of them correct? Would you rather be doing something more fun instead? No, wait, there isn't anything more fun – statistics is the most fun anyone can have by themselves…

In statistics, confidence isn't about your ego, it is a calculation of how sure you are that the result represents the 'true' measure – that is, the answer you would get if you measured the whole population. If you repeat the height measurement experiment lots of times, you will get many different answers. Most will be quite similar and a few will be quite different, but each of the answers gives us more information about the true measure. Are all the results quite similar or is there a lot of variation?

Take the middle half of the results by ranking all the results from smallest to largest and discarding the upper and lower quarters of all the results. If these results fall between 150cm and 200cm, how confident will you be in stating that the mean height is 175cm? Not very. OK, so let us instead say that the middle half of the results lie between 170cm and 180cm. Now how confident are you that the mean height is 175cm? Feeling a little more smug now, aren't we?

By now, you should be starting to get the sense that the calculated mean is just one possible measure out of many. In other words, it is not likely to be the 'true' value. In fact, you're never likely to know exactly what the true value is, and the best you can do is give some estimate of how confident you are that it lies between some lower and upper measurement. As above, you could perhaps say that you estimate with 50% confidence that the 'true' height of men lies between 170cm and 180cm (since 50% of the means lie between these values). These **confidence intervals** (170-180cm) are quite tight and the level of confidence (50%) in the estimate of the true mean is quite low; it is the same probability as tossing a coin – I don't know about you, but I'm not going to put my shirt on it. To increase confidence in the result we can widen the confidence intervals. We might then find that 75% of the results lie between 160cm and 190cm, or that 95% of the results fall in the range 150-200cm. In this latter example, we can say that 95% of experiments carried out just like this one will include the true mean, but 5% will not – there remains a 1 in 20 chance that our confidence interval does not include the true mean.

Let's have another look at the example from the top of this chapter, where the researcher pointed out that his results were twice the size of the control sample. It is very common to see this kind of graph, accompanied by a lot of excitement. The more important the discovery, the more excitement goes along with it. You can feel the researcher sensing that their career is about to take off and shoot through the stratosphere. Until the more experienced researcher coughs politely and apologises for asking a silly question, but 'umm, why aren't there any **95% Confidence Intervals** on the graph?'.

Here is a graph of the sample and control with 95% confidence intervals:

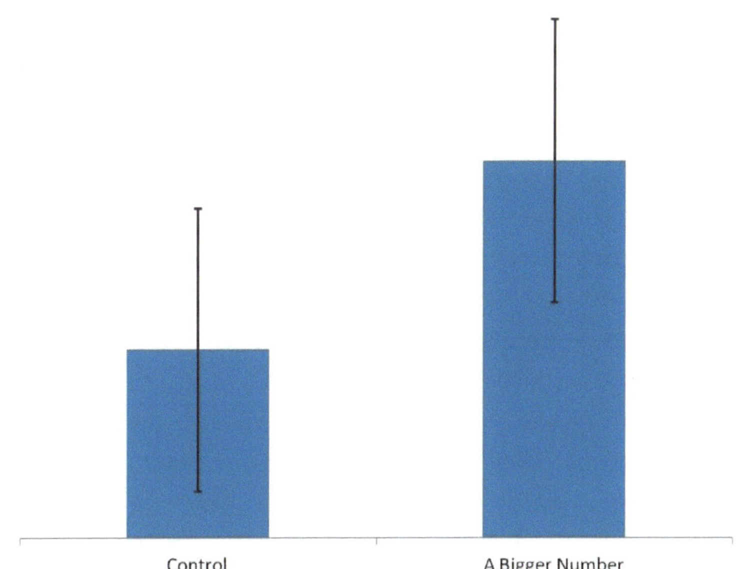

Are we still confident that we have made a major discovery? The height of the sample is still twice that of the control, but my – look how wide those confidence intervals are.

You see, the measurement of the mean is not important, but the confidence intervals are, and this is why sampling is so important to any study. With a small sample size, you can achieve pretty much any mean value you desire. The confidence intervals will betray you, but only if you *publish* them with your study.

Remember the "9 out of 10 cats" example from earlier? Given that the company commissioning the research is likely very rich (otherwise they wouldn't be spending hundreds of thousands of pounds on a large advertising campaign), don't you think they could have afforded to do a study with more than 10 cats? Of course they could, but getting the result they desire depends on a small sample size and the likelihood that they could actually repeat their results occasionally. If they had used 100 cats, how likely do you think it would have been for them to be able to publish a result that says "90 out of 100 cats preferred MiaoMix cat food (p=0.0001, 95% confidence intervals: 87.3-92.7% in over 1000 repeat experiments)"? Believe me – if these were their results they would be shouting them from the rooftops, and you would be left in absolutely no doubt that their cat food is the best. These are not their results though. They state '9 out of 10' because they cannot get 90 out of 100. There is no p-value because their results are not significant. They make no mention of the repeatability of their results because their actual results vary extremely widely. They don't publish their confidence intervals because that would actually have the opposite effect – cat owners would be disgusted and would switch to another brand. Telling the truth in a commercial setting can be the equivalent of business suicide!

A few years ago, I was attending a business course and I was chatting with the presenter at the break. Unusually, he became very interested when I told him that I was a statistician – they generally begin looking for the nearest available exit.

Incidentally, that reminds me of a story that my old professor told me and my colleagues when I was a physics undergraduate student. He said that if we are at a party and are asked what we do for a living, under no circumstances should we say that we are physicists. If we have any designs on settling down with a wife and having 2.3 kids, or even just settling for an occasional girlfriend or being invited to any other party, we should say that we are lorry drivers or postmen. I must say that it is pretty good advice, and I get invited to at least two parties most years. Now I'm a statistician I tell the same story to my students, although of course I substitute 'physics' for 'statistics'. But I digress – back to the business course.

The presenter, a former Chief Financial Officer of a very successful FTSE 100 tech company (before they were acquired by a much larger Japanese rival), told me the story of how the company spent millions of pounds trying to figure out why the reported accuracy of their device was lower than that of their main rival. They looked at every component, all the hardware, the software, the testing methods – everything. They were convinced that every component in their device was at least as good as the equivalent component in their rival's device, and in many cases superior. And yet their numbers didn't tally. They finally turned to a statistician to lead the investigation. He found that, although the reported accuracy of their device *appeared* worse than that of their rival's, the confidence intervals of the competing devices were, in fact, pretty much identical – their device was not inferior to their rival's product after all!

Now if only they had consulted a statistician in the first place. After all, we are pretty easy to spot – we're the uncomfortable looking guys at parties claiming to be lorry drivers and postmen...

Pay No Attention To That Man Behind The Curtain…

The British illusionist Derren Brown once famously claimed that he could flip a coin fairly, under controlled conditions and – in front of multiple cameras recording continuously from multiple angles – have it come up heads ten times in a row. Then he did it. You can imagine him putting his fingers to his temples, saying "now I want all of you at home to focus…", ignoring the fact that the footage had been pre-recorded and the audience was being asked to mentally influence an event *that had already taken place some weeks earlier*. "Focus…" he might have said, "and I want you to picture in your mind the image of heads coming up each and every time". "Focus…" he might have repeated, because he didn't actually say any of this, I'm using artistic licence to raise the tension for dramatic effect, "…focus, and visualise a head coming up with each new coin toss…".

"Are we ready? Let's count them off together," he says, as he tosses the first coin into the bowl.

"1 – that's the first head", he says as he picks up the coin and prepares for the second toss.

"2, that's 2, that's the second head"

"3. That's 3 heads"

"4. That's 4 heads in a row". *Sniff*, he sniffed nonchalantly

…

…

"8. That's 8 now, just 2 more to go". A little rub of the nose and a shrug of the shoulders.

"9 heads. Last one coming up", he said confidently.

"10. That's 10 heads in a row. Thank you very much ladies and gentlemen. Good night, and don't forget – I'll be here all week…"

Ta-daaaaa! Ten heads in a row. It's a modern day miracle!

Then he revealed how he had done it.

The likelihood of ten heads coming up in a row is about one in 1000 (you can do the maths if you'd like), but if you flip enough coins you will eventually get those ten heads that you are looking for. Start from the beginning, with the camera rolling (not live, of course – we're pre-recording this), and count the heads:

"1, 2, oops, a tails. Start again. 1, 2, 3, oh bugger, another tails. Start again…" and so on and so forth, etc., etc., *ad nauseum* until you eventually get the ten heads in a row. Once you have this footage, simply edit the film to show only the last ten coin flips, conveniently ignoring the previous 963 flips before the successful sequence emerged.

In his explanation, after several hours of flipping coins and looking at nothing but the bottom of the bowl, he is heard to exclaim, "I can't see the bowl anymore!". The filming lasted for over nine hours until he got the run of ten heads he was seeking. No doubt he ended up with a crippling migraine and woke up screaming "Tails!" in the middle of the night.

You see, Mr Derren Brown, magician, illusionist, con man, told no lies. He told the truth, but he did not tell you the *whole* truth. He used the camera to **lie by omission**, showing you only the things he wanted you to see, while leaving out other inconvenient truths. He left it to your imagination to either deduce the truth, which any self-respecting mathematician or statistician would have done, or to be truly astounded by the amazing spectacle. For the record, your author wasn't fooled – I knew what was going on. I guess you can fool some of the people some of the time…

A famous example of lying by omission came by way of the then President of the United States, Bill Clinton. When asked of his infamous affair with a White House intern, Monica Lewinsky, President Clinton answered, "there's nothing going on between us". In his grand jury testimony he insisted that he had answered this question truthfully because at the time he was asked there was no ongoing relationship with Miss Lewinsky – it had ended months previously. There was then a long debate as to the meaning of the word 'is', as to whether it only pertains to the present or whether it can be construed to also include events in the past. Hilarious! I guess the interviewer should have been more careful with the question, and should have insisted on also using the word 'was' in the dialogue. Then we could have debated the meaning of that word too!

You don't necessarily need to lie to get your point across though, there are other ways. Take the example that "83% of dentists recommend ToothWhyte toothpaste". Do they indeed? Well, it must be really good stuff and better than all that other rubbish then, mustn't it? If that were truly the case, then surely all the other toothpaste companies would have tested ToothWhyte for themselves and figured out their secret ingredient. Once they've done that, then their toothpaste will be just as good. Except that ToothWhyte don't need a super-expensive secret ingredient when they have statistics on their side. All they need to do is poll some dentists (who are no more experts on different brands of commercial toothpaste than the average Josephine on the street) and ask them "Which toothpastes in this list would you recommend (choose all that apply)?". Dentists are, of course, going to recommend *all* reputable toothpastes, and many of them would likely tick the ToothWhyte box as well as some of the others. ToothWhyte would be likely to score highly; after all, that was what the study was designed to do. As a precursor question, they only need ask, "Have you heard of ToothWhyte toothpaste before?". Any dentist that answers that they have not would be excluded from the study. While ToothWhyte received 83% of recommendations, their competitors could perhaps be scoring 90-100%, putting ToothWhyte at the bottom of the list. Nevertheless, there is no need to confuse the issue by reporting all these other irrelevant results is there? All that your audience needs to know is that 83% of dentists recommend *our* lovely toothpaste. This is called **wilful misdirection**, where you selectively report only those results that make your point and bury those that do not.

The customer might want an answer to the question "is ToothWhyte the best toothpaste for protecting and whitening my teeth?". The manufacturer of said toothpaste doesn't want to answer the question though. They cannot answer 'Yes' because it isn't true, and they can't answer 'No' because they will lose money. What they do instead is point to a result that answers a *different question*, one that *looks* similar to – but is distinctly different from – the one posed. The question written at the top of the dentists' poll may well have been "which of these toothpastes do you recommend?", but the answers elicited from the structure of the poll were more closely aligned to the question "do you recognise any of these commercial brands of toothpaste?".

Do you think it likely that a *real* toothpaste company would engage in such practices as in this fictitious example? In 2007, Colgate was ordered by the Advertising Standards Authority of the UK to drop their claim that "more than 80% of dentists recommend Colgate". This was deemed to be inherently misleading, as another competitor's brand was recommended almost as much as the Colgate brand by the dentists surveyed. Colgate had surveyed dentists about which toothpastes they would recommend, and allowed them to select one or more brands. The ASA said that the claim would be understood by readers to mean that 80% of dentists recommend Colgate over and above other brands, and the remaining 20% would recommend different brands. Apparently, the survey script also informed the dentists that an independent research company was carrying out the research, which was not true. Oh dear...

Here's another wonderful example of wilful misdirection by the ex-Chancellor of the Exchequer. I don't recall the exact details, so I've made up the numbers here, but I distinctly remember the mathematical sleight of hand trick that he used to misdirect. He was delivering the budget statement live from Westminster and announcing that he was going to increase funding for some government department or other, and he proceeded to set out how much extra funding he was giving them.

He announced that he was going to give an extra £10 million per year over the next 5 years. He then stated that this gives an *additive* £150 million over the next 5 years over and above the already agreed funding. Wait, what? What just happened? How do 5 lots of 10 make 150? It didn't take long before the press jumped on this and start asking questions. The Chancellor's explanation was thus: in the first year, the department will get an extra £10 million in funding. Over the first 2 years, they will get an extra £20 million, £30 million in extra funding over the first 3 years, and so on. Cumulatively, they will get an *additive* £150 million (i.e. 10 + 20 + 30 + 40 + 50 = 150). He was repeatedly asked whether that means that an extra £150 million of new money was going to be given to the department, and he would not directly answer the question, only repeating that there would be an *additive* £150 million in funding. He never repeated the figure without using the word *additive* attached to it. What he *really* said was that he was giving an extra £50 million in funding but he wanted it to look much bigger, so he used inappropriate mathematics to get an *artificially inflated* answer and presented that rather than the true figure that he didn't want to present. Don't we just love politicians?

Lying by omission and wilful misdirection are very simple devices to avoid telling the *whole* truth or answering the question. If you use small sample sizes it is easy to repeat your test until you get the result you want, then selectively report only those results you want your audience to see. Alternatively, you can simply design your study to answer a different question to the one that seems to be asked.

Oh yes, and name-drop as often as you can. 'These data came from a Harvard study, don't you know…' Did they indeed? And were the data analysed, interpreted and published by Harvard too? Naming of reputable sources gives the *appearance* of legitimacy to a study and makes the work look genuine, even if it has been totally fudged!

Pirates Caused Global Warming

When it comes to storytelling, we have a problem.

It's not our fault though – as human beings we are hard-wired from birth to look for patterns and explain why they happen. This problem doesn't go away when we grow up though, it becomes worse the more intelligent we think we are. We convince ourselves that now we are older, wiser, *smarter*, that our conclusions are closer to the mark than when we were younger (the faster the wind blows the faster the windmill blades turn, not the other way around).

Even really smart people see a pattern and insist on putting an explanation to it, even when they don't have enough information to reach such a conclusion. They can't help it.

This is the thing about being human. We seek explanation for the events that happen around us. If something defies logic, we try to find a reason why it might make sense. If something doesn't add up, we make it up.

At times, we might whimsically moan that it always rains right after washing the car, or that the phone only seems to ring when you step into the bath. Does the act of washing the car cause it to rain? Did the phone ring *because* you got in the bath? Of course not!

This is very familiar in the world of scientific exploration. A researcher, on analysing her data discovers an unusual correlation between a pair of variables. What she *should* do is be sceptical and try to find a reason why this correlation is not plausible, but she doesn't. She tries to find a reason to support the correlation, a reason why one event caused the other, allowing her to declare a new discovery.

The bottom line is that she's biased. She can't help it. We all are – it's human nature.

Ever heard the Latin expression *Post Hoc, Ergo Propter Hoc*, meaning 'After this, therefore because of this'? It is the basis of the saying 'Correlation Does Not Imply Causation', also known in statistics as the Post Hoc Fallacy because it's a very familiar trap that we all fall into from time to time. This is the idea that when things are observed to happen in sequence, we infer that the thing that happened first must have caused the thing that happened next.

The Post Hoc Fallacy is what causes a football manager to only wear purple socks on match days. He once wore them at a match and his team won. Obviously, it was the socks that did it. Now he fears that if *doesn't* wear them to a match the team might lose. Damn those stinky purple socks (he also daren't wash them for fear of the magic pixie dust washing out).

Post Hoc is also what makes rain men indispensible to the tribe – they believed that their rain man can make it rain. Spotting the clouds brewing in the distance, the rain man dances until it pours it down. It doesn't usually take more than three or four days of dancing until the inevitable happens. "Rain man dance, water fall from sky". It's just a good job for the rain man that the Indians couldn't speak Latin, otherwise he'd have been in real trouble...

For a humorous view of the Post Hoc Fallacy, let's take a look at Pastafarianism. It's all the rage these days. Not heard of it? It's one of the newest and fastest growing religions on the block. Pastafarian Sparrowism, to give it its full title, is a 'vibrant religion that seeks to bring the Flying Spaghetti Monster's fleeting affection to all of us, through the life of His Prophet, Captain Jack Sparrow'. Seriously, they're not joking. Well, actually, they are. They promote a light-hearted view of religion and oppose the teaching of intelligent design and creationism in public schools. They also maintain that pirates are the original Pastafarians.

In an effort to illustrate that correlation does not imply causation, the founder, Bobby Henderson, presented the argument that global warming is a direct effect of the shrinking number of pirates since the 1800s, and accompanied it with this graph:

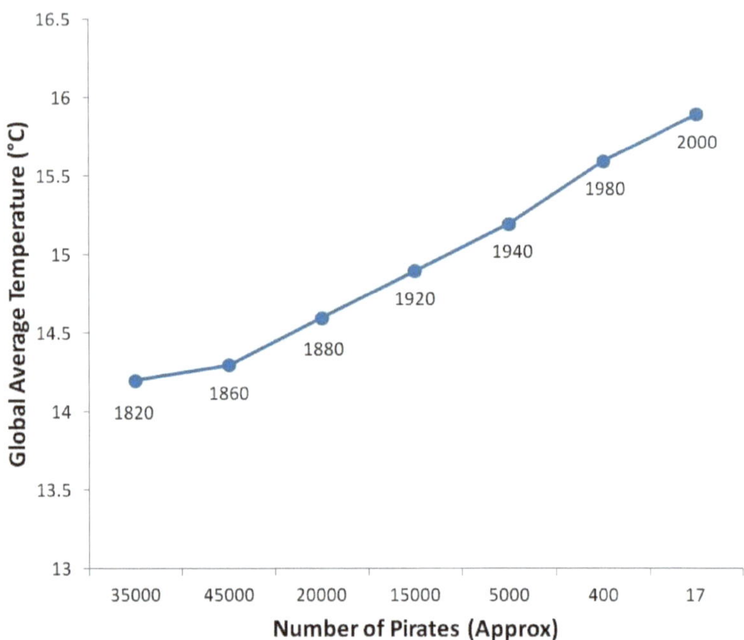

Wow, look at that straight line, I hear you all say – there's clearly a correlation between the decline in the numbers of pirates and the rise in global temperatures, so there just *must* be a causal connection here, mustn't there? Yup, you've all just fallen for the Post Hoc Fallacy (I just knew you would).

Just because there is a straight line on the graph it doesn't necessarily follow that one thing caused the other, particularly when you've grabbed two seemingly unconnected variables at random and stuck them together to see whether there might be some sort of tenuous correlation between them. In the case of pirates and global warming, take a closer look at the labels on the x-axis.

Notice something strange? Apart from the fact that the proportions of neighbouring data points are all out of whack, there is also the issue that a couple of them have been humorously disordered to deliberately deceive.

I don't know about you, but I'm a believer! As soon as I've finished writing this book I'm giving up stats for a life as a pirate on the open seas. I'll stop global warming if it's the last thing I do.

It probably will be…

If you look online there are all sorts of humorous graphs that prove the Post Hoc Fallacy. Over the past 20 years or so, there's been a huge increase in the anti-vaccine movement, particularly in the US, and there have been all sorts of spurious correlations that have been 'discovered' that 'prove' that there is a causal link between vaccination programmes and autism. We had an example of that with Dr Wakefield earlier. At the same time, to debunk the most crackpot of the theories, other – equally ridiculous – correlations have popped up too.

There was one that was published that showed the correlation between sales of organic food in the US and diagnosis of autism (next page).

There is a very close correlation between the pair of plot lines, even accompanied by a very large r-value (close to 1) and a very small p-value (close to 0). The suggestion is that – if we trust that correlation *does* imply causation – a much closer correlation exists between organic food and autism than any other theory that currently exists, so therefore it *must* be the cause. Except that correlation *does not* necessarily imply causation, and organic food does not cause autism. That would be ridiculous. And that is the whole point of these graphs. All you need to do is find any pair of variables that increase over the same time period, plot them on a graph with the same x-axis and *different* y-axes, adjust the y-axis scales until the plot lines coalesce, and – BOOM – correlation! If, by some magic of coincidence and fate, there is a *statistical* correlation, then publish the p-value that goes along with it as additional proof. What this does is prove that the correlation exists, but it does not prove that one thing causes the other. It might, but then again it might not...

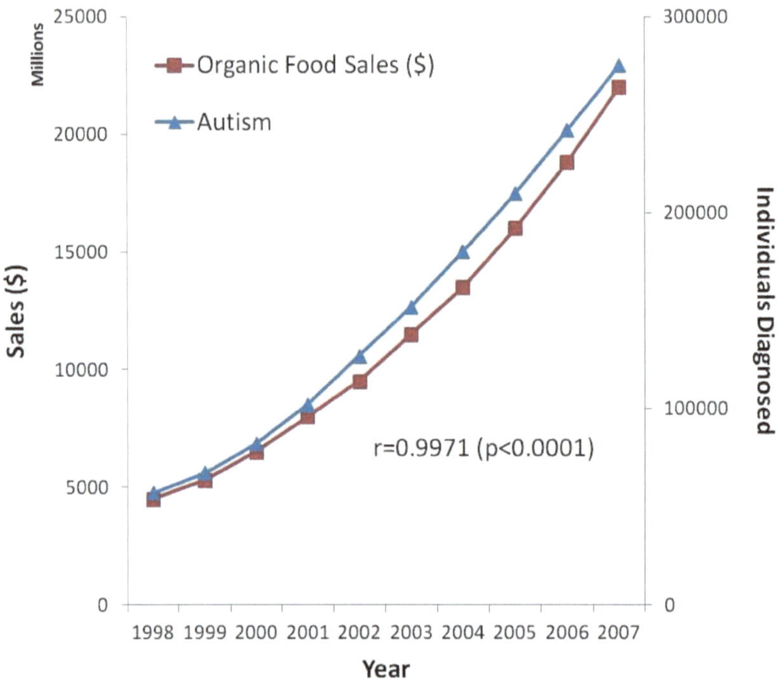

I also quite enjoyed the correlation that proved that Mexican lemons are a major cause of deaths on US roads. Wait, what? I must have missed the news that day – Mexican lemons are killing Americans? You bet!

Take a look at a plot of the number of fresh lemons imported into the USA from Mexico *versus* the total fatality rate on US highways between 1996 and 2000:

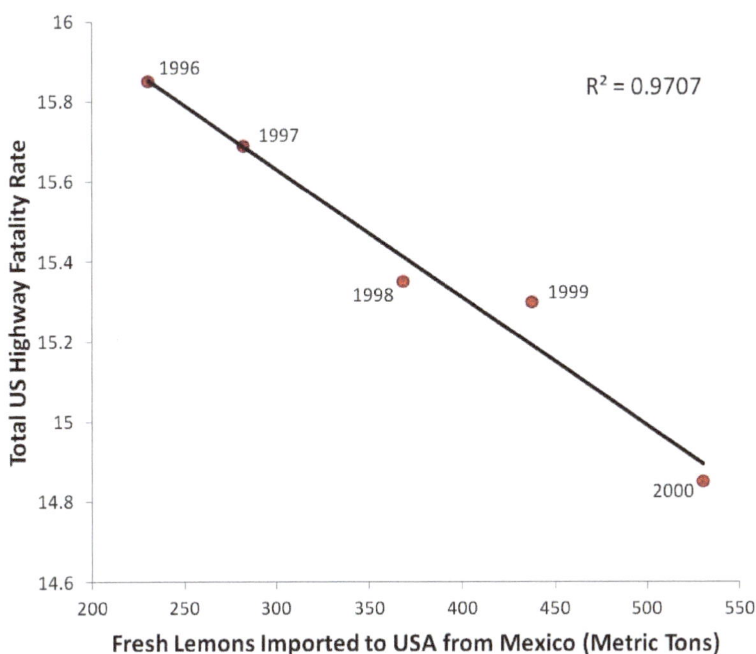

My, my, just look at the R^2 value – it really *must* be true. Although the graph seems to be telling us that the more Mexican lemons there are in the US the fewer road deaths there are, the inescapable conclusion is that MEXICAN LEMONS KILL AMERICANS! What should we do about it? Should we import more Mexican lemons (the correlation tells us that this is what we should do)? Or should we ban Mexican lemons altogether? After all, if there are no Mexican lemons on the streets then they can't kill any more Americans.

What utter tosh! I don't care if there *is* a correlation, there is nothing to suggest that lemons cause accidents. If there was, don't you think that lemons would be causing accidents on Mexican roads before the trucks crossed into the US? What about Sicilian lemons? Do they cause road deaths in Italy and across Europe?

Oh, the power of correlations. As long as your audience doesn't understand that correlation does not necessarily imply causation you can make them believe pretty much anything.

Like Microsoft Internet Explorer has been a silent killer for years. You don't believe me? Just plot the market share of Internet Explorer against the US murder rate between 2006 and 2011, like this:

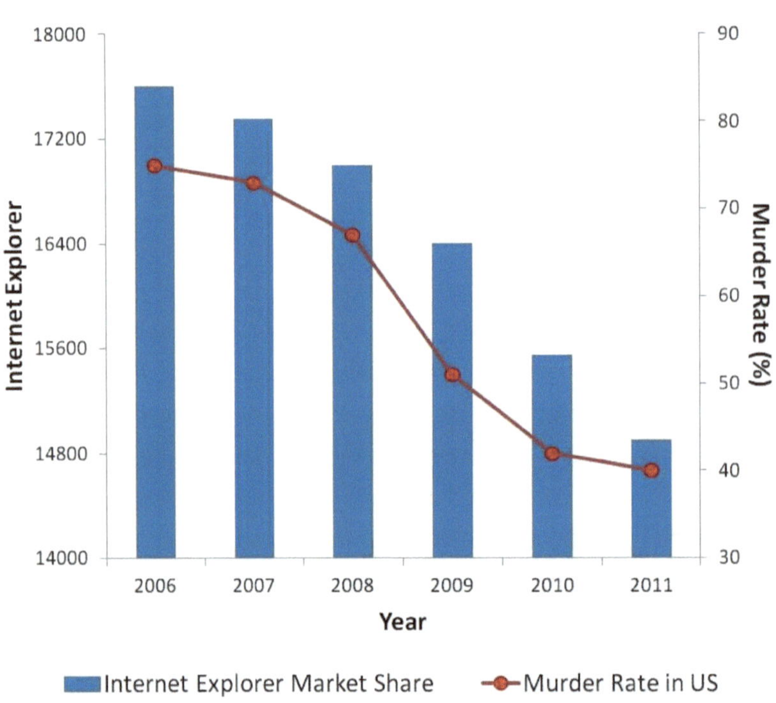

Clearly there's something insidious that lies within Microsoft's IE source code. Maybe there are subliminal messages boring into our consciousness telling us to kill, Kill, KILL!

Rubbish!

Correlation and causation are sometimes connected, but just because a graph says, 'psstt, look over here', winking and furtively raising its eyebrows, it doesn't mean that we should be fooled.

And just remember – everyone who has ever eaten an apple has either died, or will die in the future. Think about that the next time you give your child an apple…

The End (Nearly)

I hope that by now you've figured out that, if you have a mind to, you can make numbers, data and statistics say whatever you want them to say.

You can be biased in the way you select your data, including or excluding particular subsets that more easily allow you to make your point. Like running a poll in the Sun newspaper (a self-confessed supporter of the Conservative party) asking readers which way they will vote in the upcoming election. Since most readers of the Sun will likely be Conservative voters, you will likely get a dataset that is skewed towards the Conservative party.

Alternatively, you may bias your measurement methods to swing the needle of truth in your favour. You can run the same poll in the Daily Mirror newspaper too (a supporter of the Labour party). Only, run your Sun poll on page 3 right next to the scantily clad ladies that are known to adorn that particular page, whilst in the Mirror you can run your poll as a one square inch advertisement beneath the obituary column on page 28. This would be the equivalent of 'juicing' your weighing scales – you have one set of scales for Conservative voters and another set for Labour voters, and they are each calibrated quite differently. Simply add together the answers gleaned from each of these polls into a single dataset that is deliberately biased towards the Conservatives.

Finally, you can cherry pick by showing only those results you wish your ~~victims~~ audience to see. If you're a Labour supporter and you wish to show that Labour has broad support, but you find that your dataset shows the opposite, then dig deeper. You might just find that the Labour party has higher supporter levels among university students or more generally in the 20-25 year age band. Publish that then. You don't need to show your adoring public that you don't have support in all the other age brackets, do you?

Averages are a favoured way, particularly in the press and in politics, of swaying opinion towards your way of viewing the world. Simply by knowing which of the dozen or so different measures of centrality are likely to be biased in your favour, you can make the calculation say what you wish.

In the UK, there is a lot of pressure on energy companies to come clean about their profits and to charge their customers lower prices for the gas and electricity they consume. On the other hand, there is also pressure on the same companies from their shareholders to make as much profit as possible. How can an energy company satisfy both the stock exchange and the regulator, given that they have opposite requirements?

Using averages is one of their tools of choice. To the regulator they may promote that 'over the last year an extra 1 million households have saved £100 *on average* on their annual bill'. Hmm, and which measure of average would that be then? And what about the other 15 million households that you provide gas and electricity to? What happened to their bills? I don't know about you, but my annual gas and electricity bill has never gone down. Ever. Nevertheless, it's a line like this that will keep the regulator at bay for another year.

Now to make the investors happy. 'Despite fierce competition from our rivals' they might say, 'we have worked tirelessly to reduce *average* wholesale prices for gas and electricity, hence why we have made higher profits than at any time in the past 50 years…'. It's nothing short of a commercial miracle – an energy provider that can make record profits year after year and apparently still keep reducing prices to its customers. Oh, the power of averages…

And the most sacred way of deception in the commercial world is the use of small sample sizes, where anything is possible. You can make 9 out of 10 cats go to the 'right' bowl of cat food easily enough if you repeat the experiment enough times, and when you do, you have a result that you can take to the bank. If you have inadvertently dropped upon a fair, honest and truthful trial then you'll have p-values and confidence intervals to back you up and prove to your adoring public that you're not trying to pull a fast one. On the other hand, if your p-values and confidence intervals point the other way, well, that's not a problem – just don't publish them. What your fans don't know won't hurt them.

But what should you do when you can't get 9 out of 10 cats to go to *your* bowl? Well, just change the question. When you publish the results to your study you'll be open and honest about what the answers were, but you won't necessarily say what the question was. This was just what Colgate did when they published the results of a study, proudly stating, "more than 80% of dentists recommend Colgate". The answer was clear and honest, but the question was not – they designed their study so that it would more likely answer the question 'which of these toothpastes do you recognise?' rather than 'which of these toothpastes is best for clean and healthy teeth?'.

Finally, if all else fails, then you still have correlations on your side. Of course, *you* know that correlation does not necessarily imply causation, but does your doting audience know it too? If not, it is very easy to show a correlation between two things that both just happen to be rising over a selected period and suggest an explanation as to why one causes the other. Mexican lemons correlate with deaths on American roads, therefore Mexican lemons (but not lemons from Panama or Colombia) cause accidents on American roads (but not on Mexican roads). Pirates caused global warming and organic food causes autism.

Science causes suicides, as does becoming a lawyer in North Carolina (yes, these are real correlations). The consumption of cheese causes death by becoming tangled in bed sheets, as does the revenue generated by skiing facilities (these are real correlations too). Margarine causes couples in Maine (but not in New Hampshire, Vermont or Massachusetts) to get divorced, and the marriage rate in Kentucky (but not in Ohio, Indiana or Illinois) also seems to cause people to drown by falling out of fishing boats.

Coincidence? I think not!

Part 2:
Graphs Don't Lie

or

How to Lie with Graphs and Get Away With It...

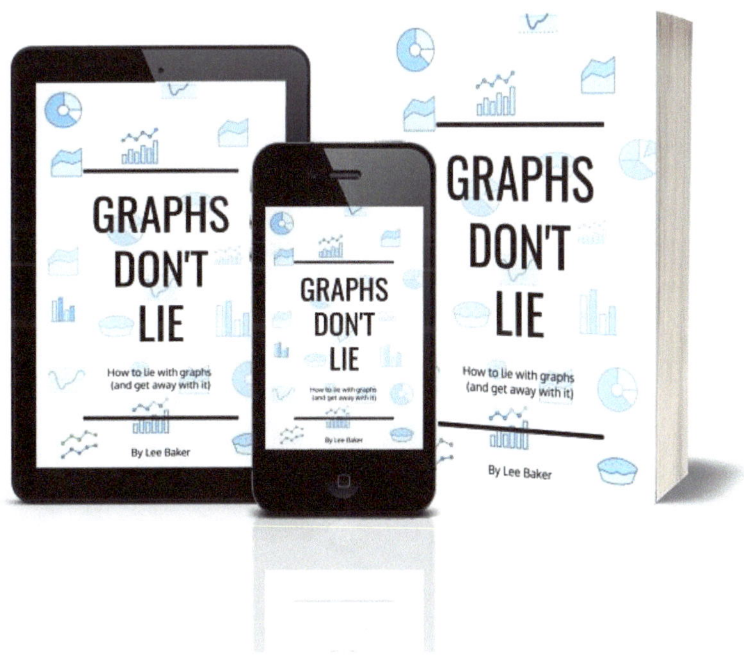

Contents

Introduction

When I do something, I give it everything I've got. It's just the way I am and I don't think I'll ever change. And if you add up all of that 'everything', what do you get? 100%. You can't give more than that. Unless you're a footballer, where 110% is standard. But for the rest of us mere mortals, it's 100%.

Apparently, though, if you work for Fox News then 193% is the sum total of everything there is.

During the 2012 US primaries, Fox News presented a wonderful pie chart showing the support levels that all three candidates enjoyed. It was a very slick presentation with a nice, shiny 3D chart, and anchorman Bryon Harlan said that Sarah Palin currently has the most support for the 2012 presidential nomination among Republican candidates.

"When it comes to landing the nomination," said Harlan, quoting directly from the pie chart on the screen, "Palin is at 70%, about a third higher than this past July". Wow, 70% – it looks like she's running away with it. Surely, none of the other candidates could catch her now, not with such large levels of support.

"Mike Huckabee," he continued, "stands at 63%, with Mitt Romney at 60%". Wait, what? What just happened? The total support for all three candidates apparently stands at 193%. What does this mean? Does it mean that 193% of eligible voters will turn out in the primaries, or perhaps that this pie chart is 193% as effective as typical pie charts? No idea, but what I do know for sure is that there were some seriously under-handed shenanigans going on there.

You see, the thing about graphs is that they don't lie. You put some numbers into a spreadsheet, tell it which graph you want, and out pops a faithful representation of your data. If you give it the numbers 70, 63 and 60, it will plot them on your chosen graph, even as a pie chart. It won't change the numbers, give them political spin or otherwise misrepresent them. It doesn't know how to.

People, on the other hand, do.

Graphs might not lie, but if you plot biased data on a graph or elect to represent truthful data on an inappropriate graph, you'll be guilty of misleading your audience. And that's the whole point.

When plotting graphs and charts you have choices and some of the choices are more appropriate than others. Do you want to tell the truth? Then choose wisely. Would you like your company to make more money or your preferred candidate to garner more votes? Perhaps you might prefer to choose a little less wisely. If you're comfortable with using graphs to misrepresent the truth – and even prefer them as a vehicle for profit or career enhancement – then this book will help you find lots of ways to tell untruths with your graphs.

A word of caution though: this book was not written for you. It was written to help the conscientious spot when you're trying to pull the wool over their eyes. After all, the crooks and conmen already know these tricks, and it falls to honest men to learn them in self-defence.

Truncate the y-axis

Most graphs you'll come across are at least mostly truthful, but from time to time, you'll come across the odd one here and there that just makes your eyes pop out, like this one:

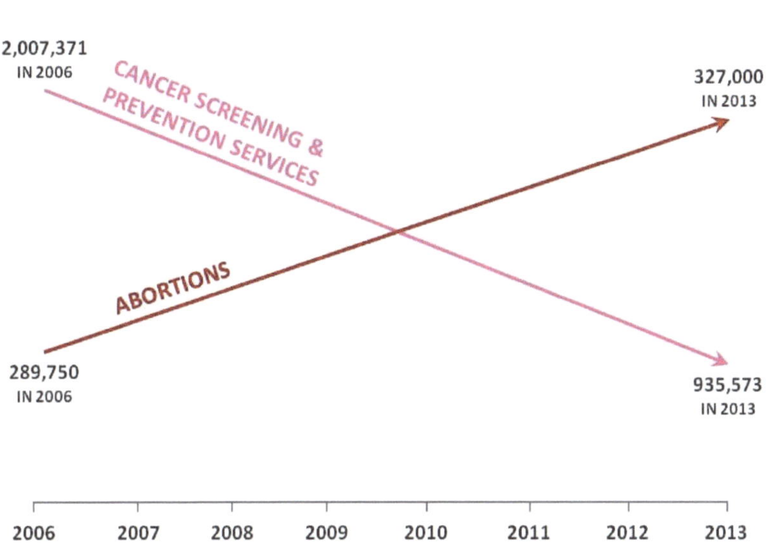

Referring to this graph in a 2015 Congressional hearing, Republican Jason Chaffetz said "In 2006, Planned Parenthood performed more prevention services and cancer screenings than abortions, but in 2013, there were more abortions".

"In pink," he continued, "that's the reduction in the breast exams, and the red is the increase in the abortions. That's what's going on in your organisation."

Oh my, where to start...

Firstly, the chart features a pair of lines crossing, one heading upwards (abortions) and the other heading downwards (cancer screening and prevention services). The data appear to show substantial growth in the number of abortions, coinciding with a visually comparable decrease in the number of cancer screenings. The numbers on the chart are based on real data and at first inspection, the facts appear damning – the implication from the graph is that there has been a switch in focus away from cancer towards more abortions. We need to look closer though, and in particular notice how the two lines are on completely different scales – the cancer data are plotted in the millions, while the abortion data are plotted in the hundreds of thousands. The number of cancer screening and prevention services in 2013 is lower on the graph than the number of abortions, and yet there are almost 3 times as many cancer interventions as abortions in that year. If you believe that this graph is correct, then you must also believe that 327,000 is a larger number than 935,573!

The creator of the graph put cancer on the primary y-axis, abortions on a secondary y-axis, truncated them both so that the lines cross in the middle, then omitted both y-axes to cover up the deception. In fact checking the graph, it was discovered that the chart came from Americans United for Life – an anti-abortion group – who insisted that the graphic was "accurate and honest". Sure it was! "That graphic is a damn lie," said Albert Cairo, a visual communications expert at the University of Miami, "regardless of whatever people think of this issue, this distortion is ethically wrong." Enrico Bertini, a professor at New York University who studies data visualisation called the chart "scandalous".

For balance, on the next page are the actual data when plotted on a single non-truncated y-axis.

These data don't look quite so damning, do they? You could hardly have come to the conclusion from this graph that there was a switch in focus from cancer to abortions – in fact, it appears that there is no increase in abortions at all. This is because of the difference in scales between the two plot lines – any change in the abortion data has to be large for it to show up on the graph.

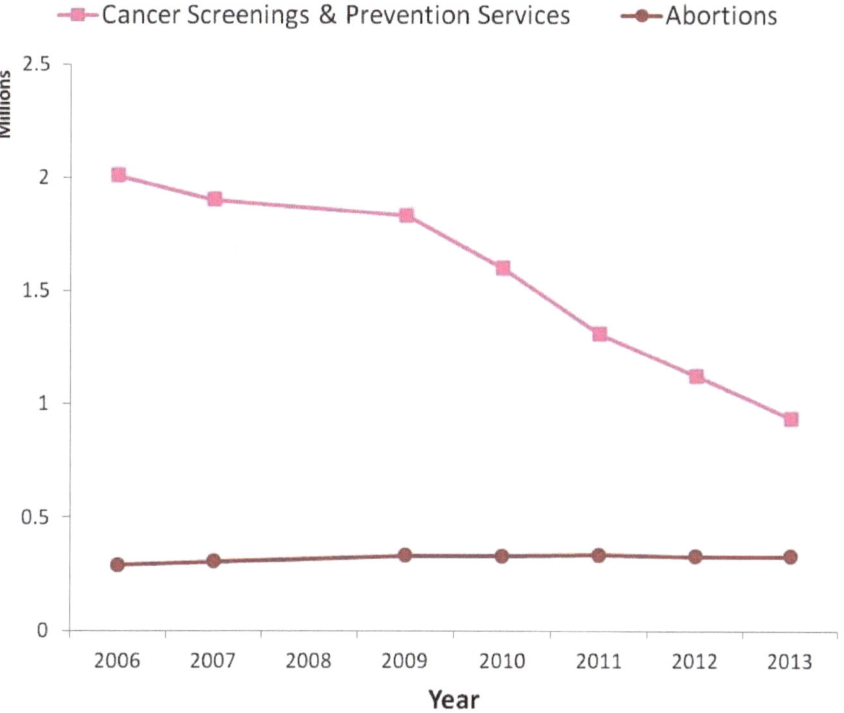

If you wanted to show the relative change of each of these two sets of data, you could plot the percentage change of each of the parameters from their 2006 baseline. Doing so would result in the graph overleaf.

From this plot, we can see that the number of abortions did increase by about 15% from 2006 to 2009. It is indeed *after* 2009 – when the abortion rate remained constant – that we see large decreases in cancer screenings and prevention services. It is clear from this graph that there is not a switch of focus from cancer to abortions, in direct contradiction to what the Americans United for Life graph appears to show.

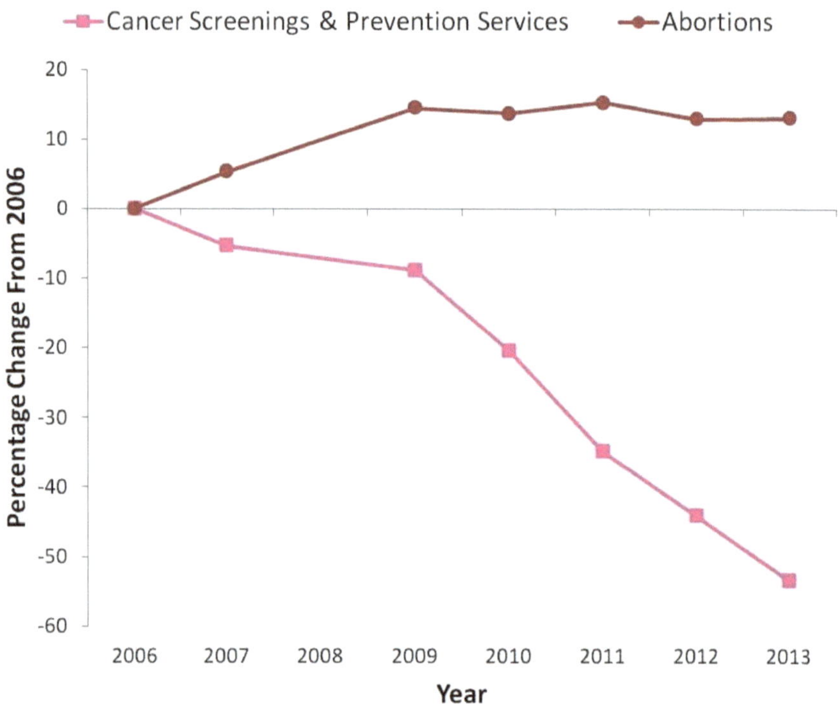

Why did the cancer screening rate drop after this year? In 2009, the US Preventative Services Task Force updated its recommendation to get mammogram screenings every two years instead of its previous suggestion of every 12 months.

Truncate the x-axis

A favoured way of misleading the reader is to show only the data you wish them to see. In statistical terms, this would be known as **selective reporting**, but we also do this in graphs by truncating the x-axis and showing only selective data that make your point.

At some point at the beginning of the 21st Century, there was a seismic shift in political opinion. Prior to this point politicians argued that the world was not warming, and even if it were, it would cost HUGE amounts of money to reverse. Then the shift happened. Politicians realised that rather than taking such costs out of government budgets they could make the people pay for it instead. Once they realised that climate change could even be *profitable*, global warming started to become something of a religion. Actually, it became two distinct religions – those in favour of the theory of anthropomorphic climate change (humans caused the planet to warm) and those against. Both sides have used precisely the same data to try to make their point.

There is one graph in particular that is used time and again by the naysayers to 'prove' that the planet is not warming, but is actually *cooling*.

Let's take a look at the data and see how they managed to come up with that conclusion. We start by taking the global mean temperature from 1880 to 2014 and graph it (overleaf).

It's not very informative – it seems to suggest that there has been no change in temperature over the past 100+ years. That's because the graph lacks context – the absolute temperature is unimportant, it is the *change* in temperature that is a cause for concern. Scientists generally agree that a change of only a few degrees would have catastrophic consequences for all life on earth, and that efforts should be made to limit global warming to less than 2.0°C (3.6°F) of pre-industrial levels.

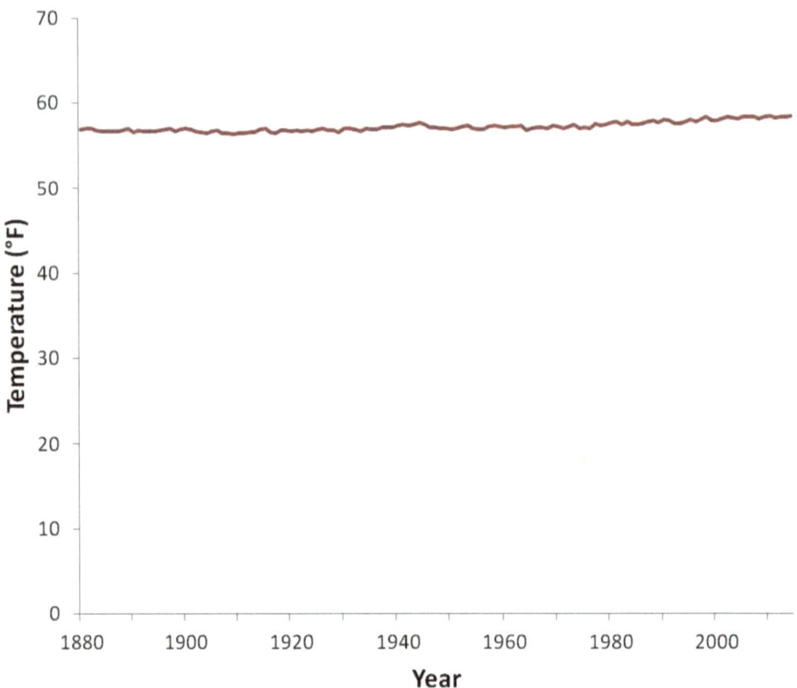

It makes sense, then, to calculate the change in temperature and plot that instead (overleaf).

That's better, much more detail. The data clearly show that there has been an increase in global temperature of around 1.7°F since 1880 and a rise of around 2.2°F between the lowest and highest recorded annual temperatures (56.35°F in 1909 to 58.53°F in 2014).

That is quite a substantial change, so it's pretty impossible to make these data show that the planet is cooling, isn't it? Au contraire!

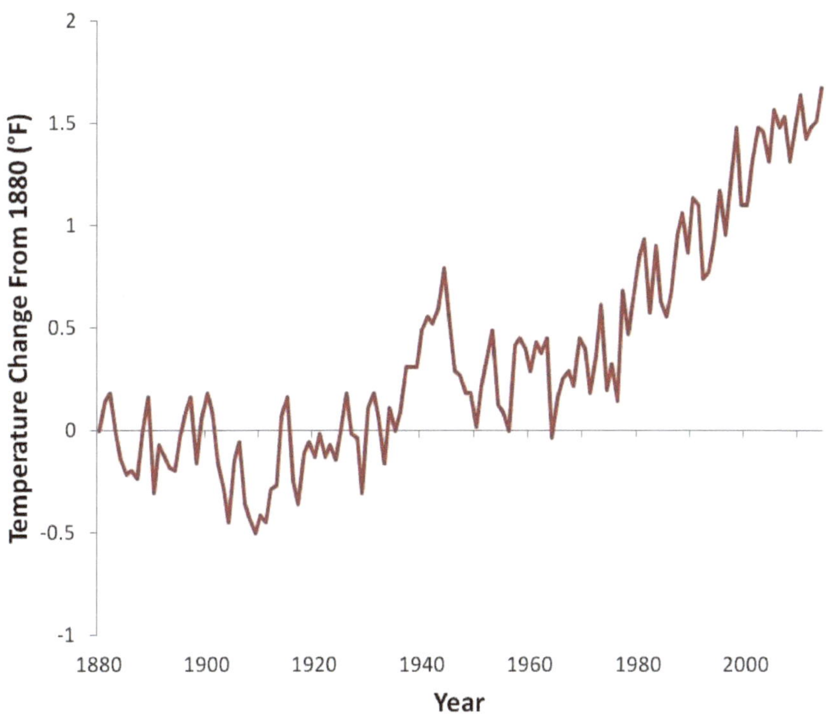

Using the simple trick of truncating the x-axis to highlight a specific range of temperatures while conveniently ignoring all other data will get us to where we need to be. Look at the region in the data between 1998 and 2011. This certainly *looks like* there could be some data in there that we can use to make a convincing argument, so let's plot that and see what it looks like (overleaf).

To aid the point we're trying to make I've changed the plot to black and added in a red arrow to accentuate the downward trend of the temperatures (as opposed to a best fit trend line, which would have an upward slope).

This graph is actually used by climate change deniers as their reference to disprove global warming. They argue that over this 13 year period there was a clear decrease in global temperatures, any warming in the planet has reached a plateau and there is no cause for alarm. Nothing to see here, move along, move along…

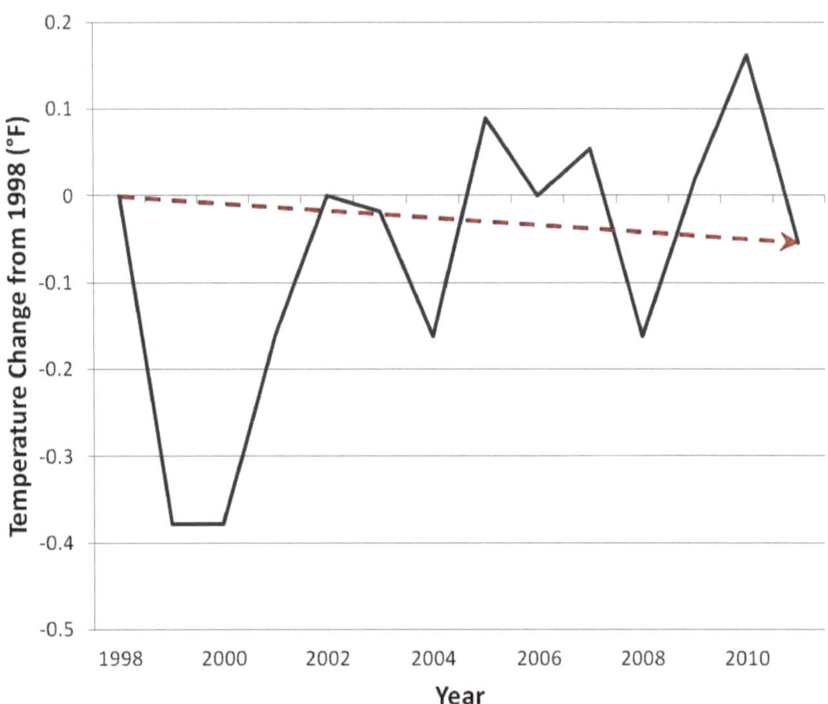

As we can see from visualising all the data though, the gradual incline of the slope from 1880 to 2014 shows that there is a clear case in favour of the theory of global warming.

Omit the y-axis labels

Back in 2013, the CEO of Apple, Tim Cook, made a live presentation to show how sales of the new iPhone had taken off. I wasn't there (I guess my invitation got lost in the post), but I have seen some photos of the event. One of them really stood out, but for all the wrong reasons. Monsieur Cook had put the following graph up on the screen to illustrate his point:

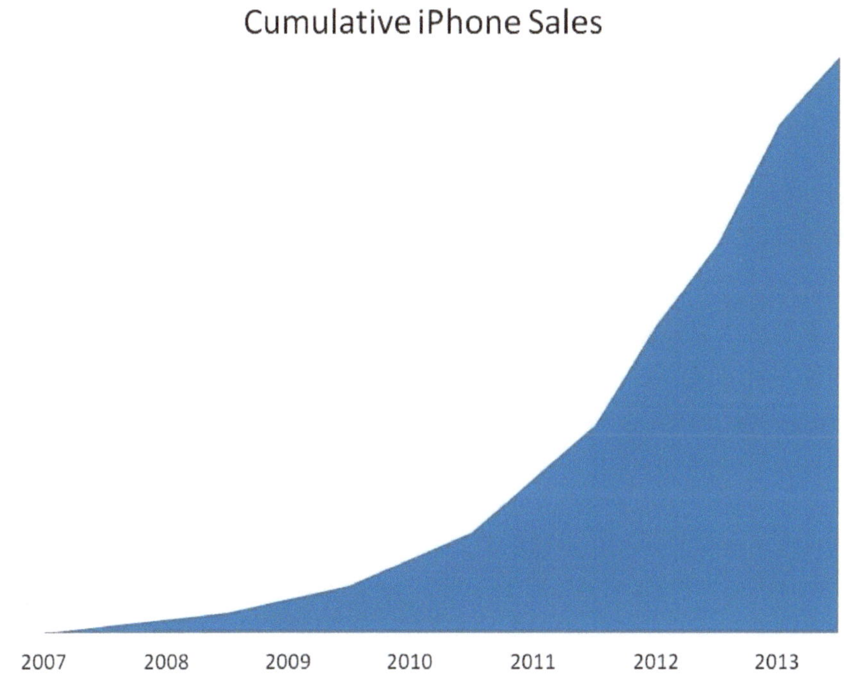

To be clear, I haven't had access to the actual data that sits behind this graph (and I bet that only a few select people in the world have), but I made measurements of each plot point with a ruler, and then recreated the graph. It's a pretty accurate representation of the actual graph that he showed, even if I do say so myself!

The point is that the graph keeps going up as you move your eyes from left to right. Sales are increasing! Or are they? Notice that there is the word 'Cumulative' in the header. That's the first clue that all is not well. If sales had indeed continued to increase year on year he would have shown it! If he's showing us cumulative sales, then he's trying to hide something. The second clue, and is meant to deceive (sorry Mr Cook, but it is, isn't it?), is that he deliberately omitted the y-axis. How many sales were there? No idea! There could have been billions of iPhone sales or there could have been hundreds. What is clear is that if there had been a lot of them, we'd have known about it. I inferred from this graph that not only have there not been as many iPhone sales as they were expecting, but that sales weren't progressing so well.

Let's deconstruct the data and see what the graph of Actual iPhone Sales looks like (see graph overleaf).

Whilst it appears to be the case that iPhone sales experienced strong growth in the earlier years, after 2012 sales seemed to drop off in some periods. In particular, the final bar on the graph is significantly shorter than each of the previous three. Is that a cause for concern for Apple's CEO? I'm not suggesting that these *aren't* healthy growth figures, but clearly, Tim Cook wasn't entirely happy with them, otherwise he would have just given it to us straight. Perhaps the total number of sales was the reason for the deception in his graph, but we will never know – the y-axis was omitted, so he could have us believe anything he wanted!

Actual iPhone Sales

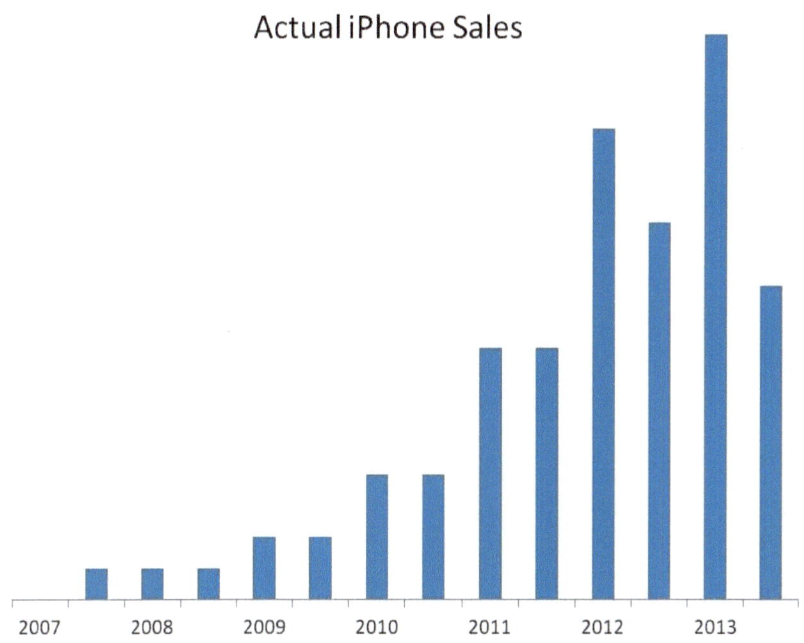

Extrapolate

Jim Hines was the first man to break the 10-second barrier for the 100 metres sprint in a time of 9.95 seconds at the Mexico Olympics on 14th October 1968. In a remarkable race, dubbed the Night of Speed, Ronny Ray Smith and Charles Greene (the previous world record holder from the night before) also ran under 10 seconds. It was to be another 15 years before the world record time was beaten again, but in a golden age of men's sprinting the record was bettered eight times (excluding drug-assisted times) in a 12 year period from 1987 to 1999. Then came Usain Bolt, who made all previous records look like they were set whilst running into a hurricane strength headwind. Oh my!

Overleaf is a graph of all the world records in the men's 100m since Jim Hines first broke the 10-second barrier.

If you had plotted these data in 1990 and looked at the graph, you would see that the plot points fall on a straight line. Extrapolating (extending into the future), you would be able to predict the progression of the world record times for the 1990s, the 2000s and 2010s. If you did that, you would see the line labelled '1' on the graph, which predicts that the 9.9-second barrier would be broken around 2010-ish. If you'd rushed off to the bookmaker's to put your shirt on it being broken in 2010 you would have been disappointed – the 9.9 second barrier was broken as early as 1991 by Carl Lewis.

This is the problem with extrapolating. It assumes that the pattern that was observed in the past will continue into the future. If there is one thing that I can guarantee in this life, it is that change is inevitable. And taxes. Change and taxes. Two things, two things I can guarantee – change and taxes. And death. Three things can I guarantee…

Oh dear, I seem to have lost myself in a Monty Python sketch… I'll come in again.

Sub-10sec World Record 100m Times (Men)

If, on the other hand, you were to consider only the golden age between 1987 and 1999 and extrapolate from there, you would get the line labelled '2'. Notice how very few of the plot points beyond 1999 actually lie on the extrapolation line... Finally, extrapolate the world record times from the Usain Bolt era and see what happens. The first person to break through the 9.0-second barrier will do so in 2010. Quick, get down to the bookies and... Oh, wait, we can't – it's already 2017 and the world record still stands at only 9.572 seconds.

We can continue doing ridiculous things like this and extrapolate line 3 even further to predict that eventually the 0.0-second barrier will be breached and the winner of the race will pass the finish line before the starting gun has been fired.

Here is what Mark Twain had to say on the subject of extrapolation:

"In the space of 176 years the Lower Mississippi has shortened itself 242 miles. That is an average of a trifle over one mile and a third per year. Therefore, any calm person, who is not blind or idiotic, can see that in the Old Oolitic Silurian Period, just a million years ago next November, the Lower Mississippi River was upwards of one million three hundred thousand miles long, and stuck out over the Gulf of Mexico like a fishing-rod. And by the same token any person can see that 742 years from now the Lower Mississippi will be only a mile and three quarters long, and Cairo and New Orleans will have joined their streets together, and be plodding comfortably along under a single mayor and a mutual board of aldermen. There is something fascinating about science. One gets such wholesale returns of conjecture out of such a trifling investment of fact."

And this very trick is commonplace to the statistical conman – in 1995, this remarkable statistic was quoted in a reputable journal:

"Every year since 1950, the number of children gunned down has doubled".

Do you want to do the maths on this or shall I?

If we start out with a single death in 1950, there will be two deaths in 1951, four in 1952, 1,024 in 1960 and over a million in 1970. We are nowhere near 1995 yet, and already the numbers are looking scary. Let's keep going. In 1980 there will be a billion children gunned down, a trillion in 1990, 35 trillion in 1995 – the year the quote was published – and a billion trillion squadrillion by 2000. Oh my, it seems the human propensity to gun down children has spread far beyond the Earth and has now infected the rest of the Galaxy. Where do we go from here and what should we do next? Get more children or fewer guns, perhaps?

Pie Charts

In Cockney rhyming slang, 'Pork Pies' means Lies. That's quite insightful, as it happens, because pie charts are one of the easiest ways of telling 'porkies', or lies. There are many ways of manipulating pie charts to make the data appear a certain way, but even with just a simple non-adulterated pie chart it can be difficult to understand what the data are trying to tell you.

Take a look at the following pie chart of the top 10 leading causes of death globally between 2000 and 2015, according to the World Health Organisation:

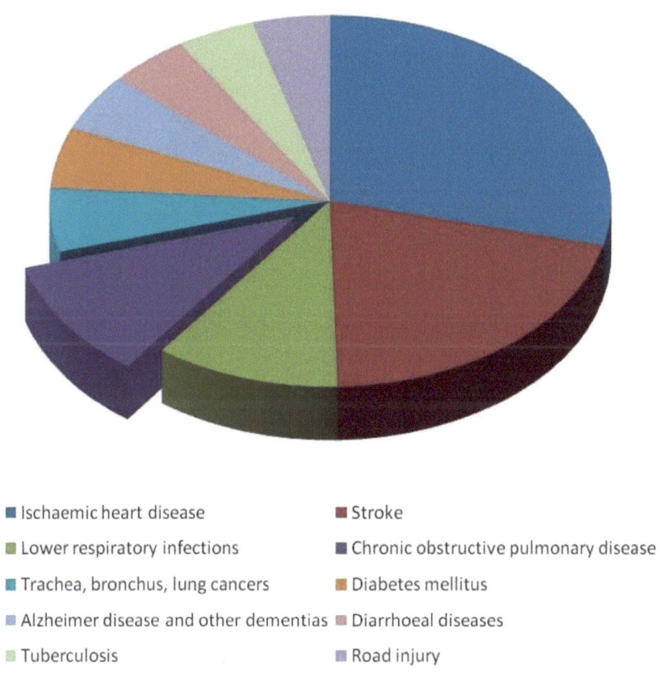

Just trying to figure out which labels correspond to the chunks of the graph is difficult, but let's take Chronic obstructive pulmonary disease (COPD) – the purple cut-out chunk (South West on the pie chart). What proportion of the whole is COPD? To my eyes, it's about 1/3 of a right angle, so about 30°. Out of 360°, that makes 1/12 of the whole or about 8.3%. How well did I do? The actual figure is 5.6%, so my estimate was too large by half.

The margins for error when making measurements and calculations of pie charts are typically quite large, which is why pie charts are used by the unscrupulous!

Comparisons are difficult to make too. Compare COPD to Trachea, bronchus and lung cancers (TBLC) – the light purple chunk located between 11 and 12 o' clock. If you wanted to compare the differences in size between these, it is difficult to make your eyes ignore the depth of the COPD section. The TBLC section of the graph has no depth, so a direct comparison can be misleading. Is COPD about 3 times the size of TBLC? Actually, it's less than twice the size.

A wonderful example of how not to present data in a pie chart came courtesy of Fox News during the 2012 primaries. The news anchor, Bryon Harlan, presented a pie chart showing the support enjoyed by Sarah Palin, compared to that of her opponents Mike Huckabee and Mitt Romney (overleaf).

The areas of the three chunks suggest that each of the candidates has roughly equal support, until you look at the percentage share of the support that each of the candidates received – Sarah Palin has 10% more support than Mitt Romney. But wait, there's more. Add up the percentages and you get 193%. What? What does that mean? Well, to get an answer to that we're going to have to dig around in the original data. It turns out that this was a telephone poll that asked 900 registered voters if they had a favourable or unfavourable opinion of a list of names. The numbers reported on the graph are the favourability ratings among voters who identified themselves as Republicans. Notice that the question in the poll ('favourability' – rate each on the list) was very different from the one reported on Fox News ('support' – choose one Republican candidate).

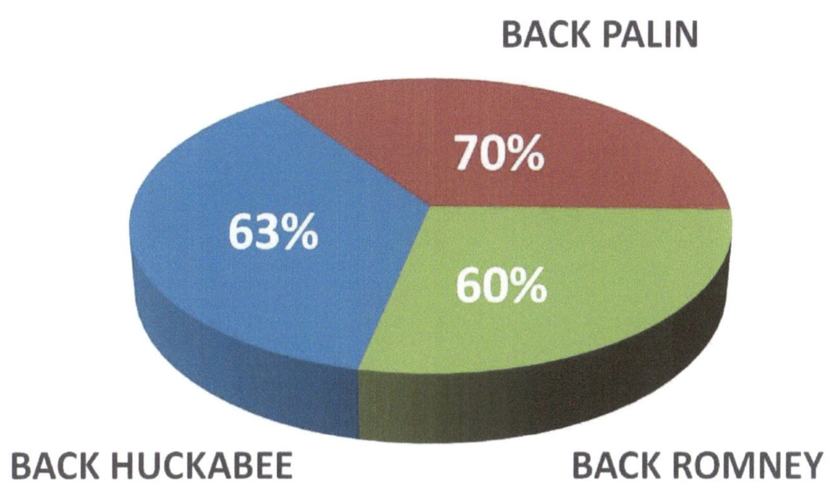

Interestingly, the original data also showed that Newt Gingrich scored 58% on the poll, just 2% behind Mitt Romney, but for some reason they decided to leave him off altogether. Baffling. Perhaps they thought it stretched credulity for the slices of the pie chart to add up to 251% rather than 193%. I guess we'll never know.

Even more interesting was that Oprah Winfrey, who wasn't a politician and wasn't running in the primaries, had a 46% favourability rating from Republican voters and an 80% rating from Democrat voters – the highest favourability rating in the entire poll.

So what can we read into these data? Pretty much nothing. Who is going to win the election? I haven't a clue, but my money is on Oprah!

Here is one of my favourite pie charts, showing the exit poll from the 2017 UK General Election, as printed widely by The Sun newspaper (and faithfully reproduced by my fair hand):

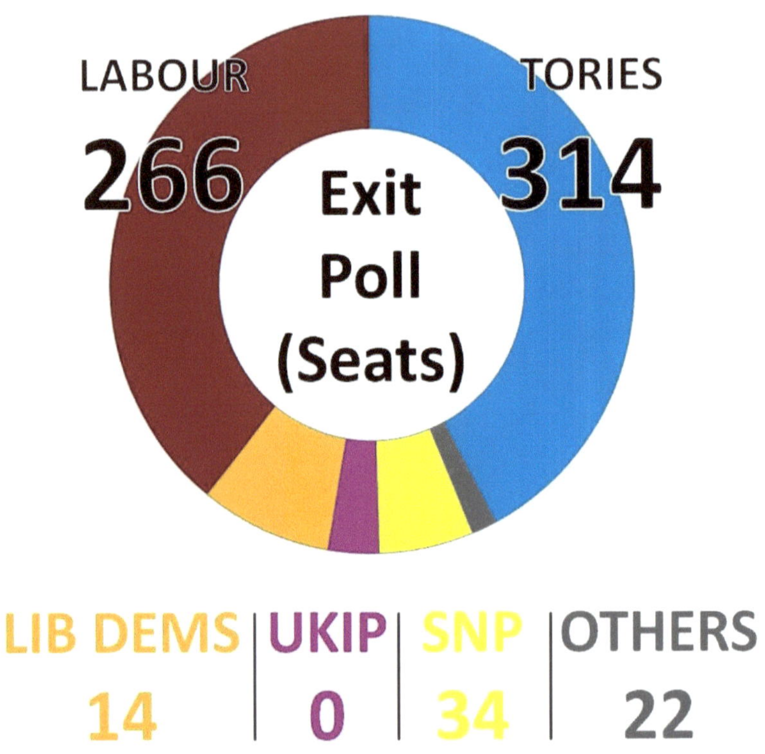

Nothing amiss here, is there? Well, nothing obvious, at least until you start to look and compare the numbers. Take a closer look and see if you can spot what is going on.

Did you notice that UKIP got no seats in the election, but they have a spot on the pie chart? Did you also notice that UKIP has a larger slice of the pie than Others, who got 22 seats?

I can imagine some kind of comedy sketch going on with these numbers, like:

"Is 34 a bigger number than 14, or is it smaller?"

"Oh, smaller, it's definitely smaller…"

"What about zero then?"

"Well, zero is a bit less than 34 but it's much more than 22…"

3D Charts

Perhaps the best way to mislead your audience with legitimate graphs is by plotting your data on 3D axes and then playing around with the various options available to you. For example, suppose that your research organisation spends far too much on administration, and although you need to report it truthfully, there is nothing to stop you reporting it unfairly. Take a look at the stacked bar chart on the left:

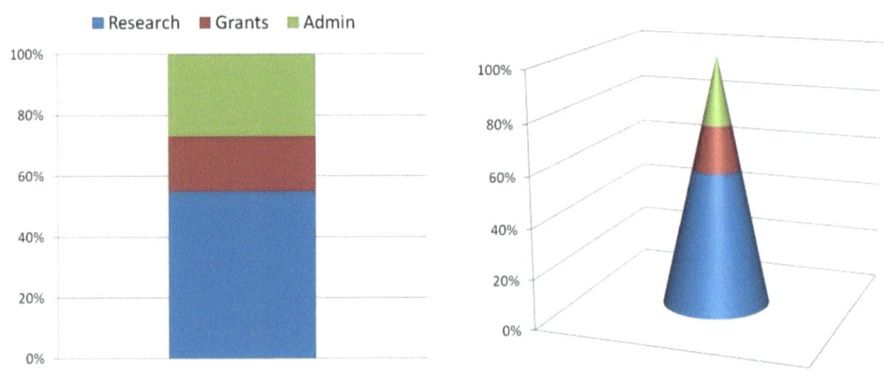

This is a fair and correct way to report that your administration costs are 27% of the total expenditure. On the other hand, 27% is quite a large chunk of all the costs and might raise questions about just what the hell you're spending it on. You need to report it correctly but make it *appear* smaller – enter the stacked cone plot on the right. With this, the correct way to read it is to take into account only the vertical height of each section. Yeah, but who does that when you have all that 3D loveliness? Of course you're going to look at the whole volume of the cone – that's why it's there! Can you really tell from the cone plot that administration costs are 50% larger than grant costs? It looks the other way around to me...

Another favoured 3D tool to play around with is perspective. Hold a toy cow between your fingers and then look at it while simultaneously looking out of the window at the real cow in yonder field. They look the same size, don't they? And yet you know that the one *out there* is waaayyy bigger than the one *in here*. That's the power of perspective – it can make even tiny things look bigger than big things. And you can use it to do the same tricks with your graphs.

Let's have another look at a couple of graphs that we've seen before and see what we can do with these when we sprinkle a little perspective pixie dust magic into the mix. Remember Tim Cook's graph about iPhone sales? Well I've re-plotted it as a 3D graph with some added perspective (below left), and voila – the slope appears much steeper suggesting a faster rate of growth, and it's also harder to spot that sales are slowing in the latter couple of years. Moreover, it would have been much harder for me to have measured the heights at each plot point to re-construct his 'real' chart!

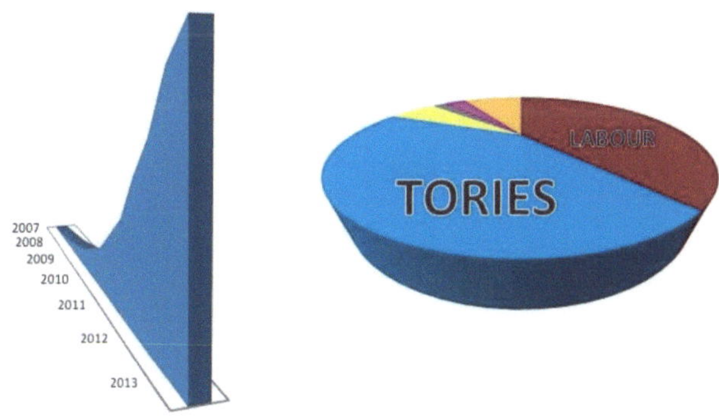

And what about the exit polls at the 2017 election? What if I had wanted to show that the Tories won by a landslide (which they didn't – they had a minority of seats and ended up in a coalition with the DUP)? There is nothing wrong with the original pie chart that a little perspective cannot fix – see the pie chart above right. Not only have I changed the chart from 2D to 3D and altered the perspective, but I've also changed the sizes of the text labels to accentuate 'Tories' over 'Labour'. In reality, the Tories won only 48% of Parliamentary seats, but on my super-duper 3D chart, it looks like they won about 60% of them. Pixie dust magic indeed!

Pictocharts

Let's face it, graphs are boring. Sorry, but they are. Whether they are line charts, histograms, time series or scatter plots, they are all somehow designed to put us to sleep. Sure, they might contain some useful information and they can be used to convey some really important facts (or untruths, as we've seen), but really, do they have to be so dull and depressing?

Well actually, no, they don't – we can jazz them up with pretty pictures, and yet still tell the same story.

Welcome to the wonderful world of the pictochart!

Whenever you see a bar chart, simply exchange the bars for colourful pictures of fruit, cows, stacks of coins, houses – in fact, anything your heart desires. All we need to do is make sure that the heights of the images match the heights of the bars we're replacing and taa-daa – instant appeal.

And they don't lie. Oh no, not pictocharts. They might give the *impression* of a different story, but interpretation is in the eye of the beholder.

Let's take an example in which we compare the average monthly wage of men in the US and the UK. Typically, we might plot these as a bar chart with, as usual, correctly labelled axes, like the one on the left overleaf:

Perfectly fine graph, you might say. All correct and proper, if a little boring and stiff. To make it altogether more exciting we could take out the bars and replace them with bags of cash, like the pictochart above right. Ah, that's better. And the information is still correct, isn't it? After all, the height of the UK wage is $2000 across both charts, and similarly $4000 for the US wage.

Yes indeed, but does the US bag of cash *look* like it is twice the size of the UK bag? Certainly not! It looks like you could get four UK bags inside one US bag, so while the bar chart image tells us that the US wage is twice that in the UK, the pictorial image conveys the message that the US wage is four times that in the UK.

Well, that is, of course, if you are assessing the area of the bags of cash. If instead you consider that the bags are three dimensional, then you'll be making a volume assessment. Then you could fit eight UK bags inside a single US bag.

This is how pictocharts are extremely misleading; persuading us that 2 x 2 does not make 4, but instead makes 8, or perhaps even 16 – and all without telling a lie!

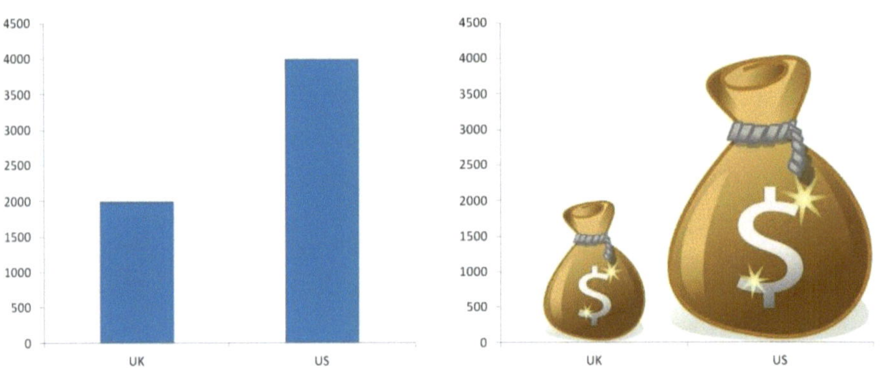

So changing the area or volumes of the image elements can make a big impact on the perception of the information contained within the data. Similarly, the image itself can shed a little shade where it ought not to. For example, let us say we wish to show the nutritional content of various fruit, or their natural sugar content or their price, or whatever. Instead of a boring histogram, we could use images of the fruit themselves as bar charts, like the one on the next page.

Clearly, three bananas are equal to three pears, and both are more than two apples. Or are we talking about bunches of bananas and pairs of pears? And did you notice how three pears are taller than three bananas? Was that intentional or are they supposed to be the same height? Now if I tell you that, buried within the text in the small print on an entirely different page, one banana represents 3000 tonnes of bananas, one apple is equal to 5000 tonnes of apples and one pear equals 2000 tonnes of pears, what would the graph now tell us? Absolutely nothing! The heights of the bar chart can't be read off the y-axis because it doesn't exist, so how could we possibly compare them? Apparently, we are meant to understand that 9000 tonnes of bananas and 6000 tonnes of pears are more than 10,000 tonnes of apples. The mind boggles!

Oh, the lies we can tell when using images instead of graphs…

Like how we can show that the decline in the population of African elephants isn't really a cause for alarm at all. We can simply show an image of an elephant to show elephant numbers in 1950 alongside a smaller image to represent the numbers in 2000, like this:

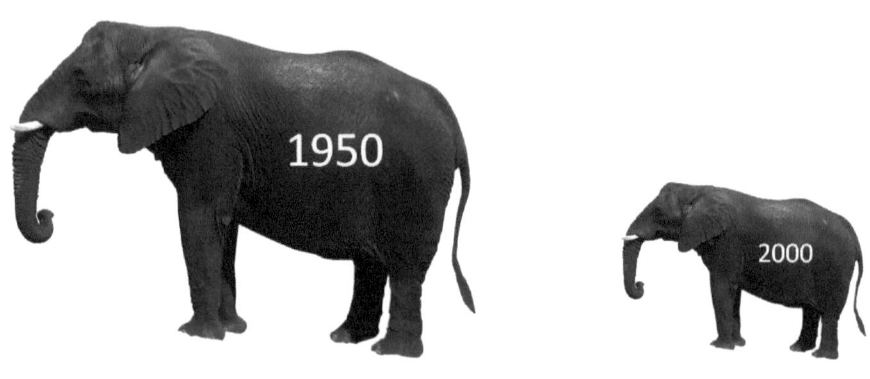

Sure, the numbers of elephants has declined, but the decline isn't so big is it? I mean, it just looks like a baby elephant with a mummy elephant. How cute! Surely, the creators of the image (the political wing of the Royal Society of International African Elephant Poachers) would be truthful and not actually misrepresent the data by fudging the sizes of the images wouldn't they?

Well, the African elephant population in the 1930s and 1940s was estimated to be around 3-5 million, and in 2014 the population was estimated to be around 700,000. Conservatively, that means that the elephant population has declined to one-quarter of its size (or worse). Reading the heights of the images (as we are supposed to do with graphs) tells us that the 2000 population is around half of that in 1950. So the graph lied then? Not quite. You see, the *areas* of the elephants are in the proportion of 4:1, so the pictograph tells the truth (ish), as long as you read the areas and not the heights. Omitting all axes from the pictograph means that the reader can interpret the image in *any way they wish*, and it is not the fault of the image creator if the reader gets it wrong...

Epilogue

When it comes to lying, there is no better way than to do it with graphs. There is something compelling about the way they are presented, the way that the 'facts' wash over you and bore into your consciousness without you being burdened by any of the tiresome details. It is easy to be seduced. Want to accentuate certain features? That's fine; you will probably fool most of the audience who won't take the time to look more closely. Would you like to sell more product or get more votes, but the numbers betray you? No problem, just grab the numbers that you wish to present (truthful or otherwise) and present them on a graph that shows your version of the 'truth' in the best light possible.

Now you know how it is done by the political conmen. When truncating axes and omitting axis labels you can distort reality to such a degree that you can make people believe that the earth is flat (yes, some still believe this) or that global warming is a fallacy (there are a *lot* of people that believe this). With extrapolation, you can convince your audience of anything you like. Any pair of numbers can be used with this trick and any trend can be up, down or in any direction that pleases you. Other plot points might give the game away, so you might wish not to show them. Remember the abortion *versus* cancer plot from earlier? How many plot points from each were used? Just a single pair. And what about the graph used to 'prove' that the world is cooling rather than warming? How many plot points were used to convince? Again, just two. See how it's done? Simple, isn't it?

At times, bar charts, line charts and scatter plots are not on your side, and somehow you can't get the data to make your point very well. That's when the humble pie chart comes in. Pie charts are about the easiest way you can find to distort the truth or even tell outright lies – and get away with it. With pie charts you can bury data that you don't want to be seen, like in the pie chart that showed the support for Palin, Huckabee and Romney – Newt Gingrich was nowhere to be seen. Alternatively, you can accentuate, like in the pie chart from the Sun newspaper that gave a spot on the pie chart to UKIP, who didn't get a single seat in Parliament. Not only should UKIP not have had a spot on the chart, but they were even given a larger slice of the pie than Others, who won 22 seats. Incredible!

The best of the numerical charts to deceive, though, has to go to 3D charts – especially the 3D pie chart. We saw how we could turn 48% of votes into what looks like 60% in the case of the 2017 UK General Election. The numbers on the pie chart were truthful (at least, for the Tories and Labour), but changing the perspective made it look very different indeed.

The gold medal for graphic deception, though, simply has to go to the pictochart. With these, you don't even need numbers. Just display a pair of differently sized pretty pictures next to each other and you can leave the rest to the viewer's imagination. As we have seen, we can make 2x2 equal 8 or even 16 without breaking into a sweat. The heights of the adjacent images may be in the ratio 1:2, but what is seen – the areas and volumes – tells a different story. It's not even your fault if the reader misunderstands it, and this is your advantage.

So now you know how to lie with graphs. Graphs themselves don't lie, they are simply a representation of the data, and it is up to you how you represent those data. Be careful though, we are just beginning a new age in the evolution of mankind – the age of data – and we are all becoming ever more aware of your con tricks.

We are watching…

About The Author

Lee Baker is an award-winning software creator that lives behind a keyboard in a darkened room. Illuminated only by the light from his monitor, he aspires to finding the light switch.

With decades of experience in science, statistics and artificial intelligence, he has a passion for telling stories with data. Despite explaining it a dozen times, his mother still doesn't understand what he does for a living.

Insisting that data analysis is much simpler than we think it is, he authors friendly, easy-to-understand books that teach the fundamentals of data analysis and statistics.

His mission is to unleash your inner data ninja!

As the CEO of Chi-Squared Innovations, one day he'd like to retire to do something simpler, like crocodile wrestling.

Leave a Review

Thank you for reading **How to Lie with Numbers, Stats & Graphs**, a book containing **Truth, Lies & Statistics** & **Graphs Don't Lie**.

I hope you enjoyed reading these books as much as I enjoyed writing them. If you did, please take a moment to leave a review. I read them all personally, and the best reviews will be featured at the beginning of future editions of the book.

- https://www.chi2innovations.com/tell-us-what-you-think-how-to-lie-with-numbers-stats-graphs/

9 798651 406098